THE COMMONWEALTH AND INTERNATIONAL LIBRARY

Joint Chairmen of the Honorary Editorial Advisory Board SIR ROBERT ROBINSON, O.M., F.R.S. London
DEAN ATHELSTAN SPILHAUS Minnesota

Publisher ROBERT MAXWELL, M.C.

STRUCTURES AND SOLID BODY MECHANICS DIVISION VOLUME 6

General Editor B. G. NEAL

structural theorems and their applications

structural theorems and their applications

B. G. NEAL
M.A., Ph.D., M.I.C.E.,
M.I.Mech.E., A.M.I.Struct.E.

Imperial College of Science and Technology, London

PERGAMON PRESS
Oxford · London · Edinburgh
Paris · Frankfurt

THE MACMILLAN COMPANY
New York

PERGAMON PRESS LTD. Headington Hill Hall, Oxford
4 & 5 Fitzroy Square, London W.1

THE MACMILLAN COMPANY 60 Fifth Avenue,
New York 11, New York

COLLIER-MACMILLAN CANADA, LTD. 132 Water Street South,
Galt, Ontario, Canada

GAUTHIER-VILLARS ED. 55 Quai des Grands-Augustins,
Paris 6

PERGAMON PRESS G.m.b.H. Kaiserstrasse 75,
Frankfurt am Main

FEDERAL PUBLICATIONS LTD. Times House,
River Valley Road, Singapore

SAMCAX BOOK SERVICES LTD. Queensway, P.O. Box 2720,
Nairobi, Kenya

Library of Congress Catalog Card No. 64-7679

Printed in Great Britain by

Set in 10 on 12 Times THE ALDEN PRESS LTD.
OXFORD

Contents

Preface viii

Chapter 1. The Problems of Structural Analysis

1.1. Introduction 1
1.2. Basic conditions 2
1.3. Exclusion of gross deformations 4
1.4. Statically determinate and indeterminate structures 8
1.5. Choice of variables 10
1.6. Direct and indirect approaches 14

Chapter 2. Principles of Superposition

2.1. Introduction 16
2.2. Superposition of force systems 16
2.3. Superposition of displacements 22
2.4. Superposition for linear elastic structures 25
2.5. Symmetry and skew-symmetry 29

Chapter 3. Virtual Work and Energy Concepts

3.1. Introduction 36
3.2. Principle of Virtual Work 37
3.3. Virtual Work transformations 42
3.4. Strain energy 50
3.5. Complementary energy 54
3.6. Total potential 57

CHAPTER 4. INDETERMINATE STRUCTURES BY THE COMPATIBILITY METHOD

4.1. Introduction 58
4.2. Truss with single redundancy 59
4.3. Truss with several redundancies 66
4.4. Proof of Engesser's Theorem of Compatibility 71
4.5. Lack of fit 74
4.6. Castigliano's Theorem of Compatibility 77
4.7. Further applications of Virtual Work 79

CHAPTER 5. CALCULATION OF DEFLECTIONS

5.1. Introduction 93
5.2. Statically determinate truss 94
5.3. Statically indeterminate truss 97
5.4. First Theorem of Complementary Energy 101
5.5. Castigliano's Theorem (Part II) 102
5.6. Further examples 104

CHAPTER 6. INDETERMINATE STRUCTURES BY THE EQUILIBRIUM METHOD

6.1. Introduction 113
6.2. Truss with two deformation variables 114
6.3. Lack of fit 123
6.4. Theorem of Minimum Potential Energy 125
6.5. Castigliano's Theorem (Part I) 131
6.6. Beams and frames 132

CHAPTER 7. RECIPROCAL THEOREMS

7.1. Introduction 141
7.2. Proofs of reciprocal theorems 142
7.3. Influence lines for redundant structures 146
7.4. Influence lines by model analysis 150

CHAPTER 8. THEOREMS OF PLASTIC ANALYSIS FOR PLANE FRAMES

8.1. Introduction 153
8.2. Plastic and incremental collapse 155

8.3. Plastic collapse theorems 164
8.4. Upper and lower bounds on λ_c 169
8.5. Shake-down and incremental collapse theorems 172
8.6. Trial and error method for determining λ_s 178

APPENDIX A. PROOF OF PRINCIPLE OF VIRTUAL WORK FOR PLANE FRAMES 183

BIBLIOGRAPHY 191

RECOMMENDED FURTHER READING 194

INDEX 195

Preface

THE response of any structure to external loads or other influences depends upon the physical properties of its members. In addition, it is necessary in a structural analysis to consider the conditions of statical equilibrium between the external loads and the internal forces in the members, and also the requirements of geometrical compatibility which relate the deflections of the structure as a whole to the deformations of its individual members. In some methods of structural analysis the conditions of equilibrium and of compatibility are determined by direct methods, that is by applying the principles of statics and of geometry, respectively. In such cases, the structural theorems are not used.

However, it may be desirable to avoid carrying out a direct geometrical analysis. The conditions of geometrical compatibility can then be derived indirectly by using Engesser's Theorem of Compatibility, or in certain cases Castigliano's Theorem of Compatibility. These theorems transform the actual geometrical problem of determining the compatibility equations into a hypothetical equilibrium problem, and are used when this latter problem is easier to solve. This aspect of the theorems is carefully emphasised in Chapter 4, where it is also shown that identical transformations may be achieved by an application of the Principle of Virtual Work.

The conditions of geometrical compatibility are also involved in the determination of the deflections of a structure, once its internal deformations are known. The deflections may also be found directly or indirectly; in the latter case the transformation of a hypothetical equilibrium problem may be brought about by the use of the First Theorem of Complementary Energy, or in certain cases by Castigliano's Theorem (Part II). Alternatively, this transformation may be made using the Principle of Virtual Work, as discussed in Chapter 5.

The conditions of statical equilibrium may be dealt with directly in an analysis by applying the principles of statics. However, it may be more convenient to derive them indirectly by using Castigliano's Theorem (Part I), or equivalently the Theorem of Minimum Potential Energy. These theorems, which are formally identical, transform the actual statical problem into a hypothetical geometrical problem, which may be solved more readily than the original statical problem. This transformation can also be achieved by the use of the Principle of Virtual Work, as explained in Chapter 6.

The Principle of Virtual Work may therefore be used to achieve both of the types of transformation which are used extensively in structural analysis, and this central role of the principle is emphasised in the book. It is also shown how applications of the principle always lead to computations which are identical with those which need to be performed when the appropriate energy theorems are used.

The two types of transformation described above are explained in detail in Chapters 4, 5 and 6. The first three chapters deal with certain essential preliminaries. Chapter 1 is a brief account of the problems of structural analysis, in Chapter 2 the various Principles of Superposition are discussed, while in Chapter 3 the Principle of Virtual Work is stated and proved, and in addition definitions are given of Strain Energy, Complementary Energy and Total Potential.

In Chapter 7 the Reciprocal Theorems of Maxwell and Betti are derived directly from the Principle of Virtual Work, and a brief account of some of their uses is given.

The last chapter of the book is devoted to the theorems of plastic and incremental collapse for framed structures, which are derived by using the Principle of Virtual Work.

The main purpose of the book is to provide a connected account of the various structural theorems, and only a sufficient number of applications is given to explain each result. These applications are generally limited to simple types of structures such as plane trusses, continuous beams and frames, since a full understanding of the principles involved may be arrived at without considering complicated structures. No attempt is made to describe all of the various techniques which are now available for the analysis of structures; these are dealt with in companion volumes in the series. For this reason, examples which can be worked by the reader are not included.

Questions of the stability of equilibrium are not discussed at all; these are also dealt with in companion volumes.

The author wishes to thank Professor E. W. Parkes, who read the manuscript with great care and made many valuable comments. He also acknowledges with gratitude the excellent work of Miss Berthoud, who patiently typed the many drafts of the manuscript.

1
The problems of structural analysis

1.1. INTRODUCTION

Structural analysis is concerned with the determination of the forces and deformations in the members of structures, together with the deflections of their joints. There are several principles and theorems which are used extensively in the analysis of structures, and the purpose of this book is to present these in a connected fashion, so that the scope and limitations of each result can be clearly understood.

It is shown that the Principle of Virtual Work occupies a central position in structural analysis; the various energy theorems which are quoted (and sometimes misquoted) in texts on this subject are all shown to be special cases of the virtual work principle. It follows that if the Principle of Virtual Work and its applications are thoroughly understood, the energy theorems need not be used at all in the analysis of structures. It is therefore recommended that the virtual work approach should be adopted in all cases, since it is then not necessary to remember all the various energy theorems and the circumstances in which they can be applied. It should, however, be emphasised that the detailed calculations involved in the solution of any particular problem will be the same whether the basic equations have been derived by virtual work or by the appropriate energy theorem.

Since the main purpose of the book is to explain the Principle of Virtual Work and the energy theorems, a full account of their applications is not necessary. In fact, for the sake of clarity, the explanatory examples which are given deal for the most part with simple plane structures, either pin-jointed braced frameworks (or trusses) loaded only at the joints, in which the members only carry axial forces, or beams and rigidly jointed frames which carry loads by virtue of the resistance of the members to flexure. Even for these two types of structure a comprehensive treatment is not given; the reader is instead referred to companion volumes in the series by Parkes (1964) and Heyman (1963).*

The problems involved in the design of structures will not be discussed. In any particular design it is necessary to ensure that the stresses in the members are such that there is an appropriate margin of safety against failure, and that the deflections are acceptably small. Design procedures are thus based on the results of structural analysis, but there is the especial difficulty that at the start of the design process the stresses and deformations throughout the structure often depend on the sections of all the members, none of which are yet known. The synthesis of a structure is in fact usually made by a procedure involving various conservative approximations, and a study of this process is beyond the scope of this book.

1.2. BASIC CONDITIONS

Forces and deformations in a structure can be caused in various ways. Externally applied loads will obviously cause internal forces in a structure, and loading due to inertia effects could be included in this category. Other possible causes of internal forces are temperature changes and the sinking of supports in the case of redundant structures. However, whatever the cause of the deformations and internal forces, three basic conditions will always need to be considered in carrying

* A Bibliography is given on p. 191.

out a full structural analysis. These are the conditions of statical equilibrium, of geometrical compatibility of deformation, and the characteristics of the individual members of the structure. Each of these conditions will now be considered in turn.

Equilibrium

The internal forces in the structure and the external loads must satisfy all the conditions of equilibrium. These conditions are usually purely statical, but occasionally there are problems in which inertia forces must also be considered.

For a statically determinate structure the internal forces by definition depend only upon the loads, and if there are no loads the internal forces must all be zero. However, for statically indeterminate or redundant structures, this is not so. For instance, internal forces and deformations may arise in redundant structures due to movements of the supports. Temperature changes can also cause stresses in unloaded redundant structures, as can the insertion of additional members which do not fit freely into place in the unstrained structure. Previous straining beyond the elastic limit may induce residual stresses which exist under zero load. In all these cases the internal forces are not determined solely by the loads, but there will be conditions of equilibrium relating them to the loads which of necessity must be obeyed.

Compatibility

The deformed members of the structure must continue to fit together, so that the deformations must be geometrically compatible. This must be true whatever the cause of deformations in the members; these can be due to stress or temperature change, and the deflections of the structure as a whole may be due to load, temperature change, sinking of supports, the initial lack of fit of members, or even previous straining beyond the elastic limit.

Member characteristics

It is always necessary in a full analysis to utilise the quantitative relationships between strain and the various agencies which produce strain; these relationships will be referred to as the member characteristics. In many cases the only cause of strain which needs to be considered is stress, and it is then the stress/strain relationship which must be taken into account. If the structure remains in the elastic range the stress/strain relationship is simple and involves a one-to-one correspondence of strain and stress. However, if some members have entered the plastic range at some stage during the loading, there is no longer a unique correspondence of stress and strain. Problems can arise in which strains due to temperature change may also need to be considered, and there are also more complex relationships such as the stress/strain/time relationships which occur in creep problems.

To summarise, structural analysis is concerned with the two basic conditions of equilibrium and compatibility. The deformations are required to be compatible, the internal forces must satisfy the requirements of equilibrium with the loads, and the deformation of each member must be correctly related to the internal forces and other strain producing agencies in accordance with the member characteristics.

1.3. EXCLUSION OF GROSS DEFORMATIONS

The deformations of structures are generally small. In many cases, they are also small in the special sense that the geometry of the structure is sensibly unaffected by deformation. This implies that the conditions of equilibrium can be written down for the undistorted structure without loss of accuracy, and further that the conditions of compatibility can also be related to the undistorted shape of the structure. The present volume is concerned exclusively with problems for which the deformations are small in this special sense.

The distinction can be made clear by the simple example of Fig. 1.1. Here a load W is supported from two fixed points A and B by a cable. It will be supposed that the value of W is

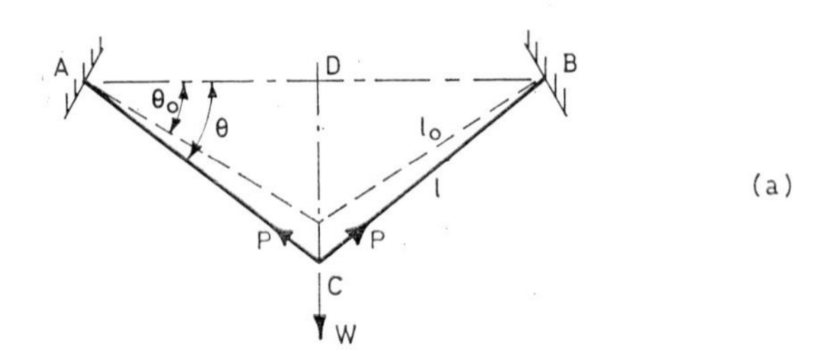

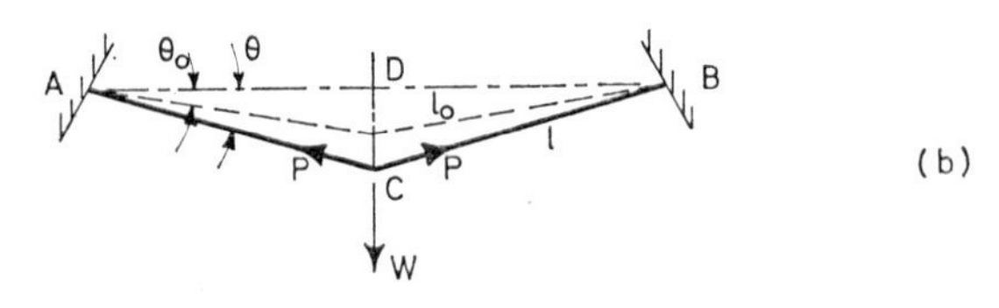

FIG 1.1

such that the tensile strain in the cable is 0·001. Then if the angle θ_0 is 30°, as in Fig. 1.1(a), and this angle becomes θ when the load is applied, geometrical compatibility requires that

$$l_0 \cos \theta_0 = l \cos \theta = \text{AD} \tag{1.1}$$

Putting $l/l_0 = 1{\cdot}001$, and $\theta_0 = 30°$, it is readily found that $\theta = 30°\,06'$.

The change in the angle θ_0 is thus small in this case. Thus if the condition of vertical equilibrium was written down for the undistorted structure, so that

$$W = 2P \sin \theta_0, \tag{1.2}$$

the inaccuracy as compared with the exact relation $W = 2P \sin \theta$ would be only 0·3 per cent.

The vertical deflection Δ of the point C may similarly be computed to a high degree of accuracy by an approximation based on the fact that the change in the inclination of AC is small. The projection of Δ on AC is approximately $\Delta \sin \theta_0$, and this is equated to the change in length $(l - l_0)$ of AC, giving

$$\Delta \sin \theta_0 = l - l_0 = 0{\cdot}001 l_0$$

$$\Delta = 0{\cdot}001 l_0 \operatorname{cosec} \theta_0 \tag{1.3}$$

The exact value of Δ is $(l \sin \theta - l_0 \sin \theta_0)$, and it is found that this differs from the approximation of equation (1.3) by only 0·15 per cent.

However, if the angle θ_0 is much smaller, as in Fig. 1.1(b), the application of a sufficient load to cause the same tensile strain of 0·001 causes appreciable changes in θ which cannot be disregarded. Thus equation (1.1) still holds, and if $\theta_0 = 2°$, it is found that $\theta = 3° \, 15'$.

With these values it would clearly be extremely inaccurate to write down the condition of vertical equilibrium for the undistorted structure, equation (1.2). This would lead to $W = 2P \sin 2° = 0{\cdot}0698P$, whereas the correct condition is $W = 2P \sin 3° \, 15' = 0{\cdot}1134P$. Moreover, the vertical deflection of C, if computed from equation (1.3), would be given as $0{\cdot}02865 l_0$, whereas the correct value is $0{\cdot}02185 l_0$.

In this latter case the deformation of the structure is said to be gross, in the sense that the loading causes significant changes in the geometry. In the former case, even though the actual member strains were the same, the deformations were small in the sense that the changes in geometry were insignificant.

Another example of a simple structure in which the changes in geometry are significant is the eccentrically loaded strut of Fig. 1.2. Here it is essential to write down the equilibrium equation as

$$W(e + y) = M, \tag{1.4}$$

where M is the bending moment at A, and y is the deflection due to W. If the relationship between bending moment and curvature is taken as the familiar $M = -EI\mathrm{d}^2y/\mathrm{d}x^2$ for elastic behaviour, equation (1.4) becomes a simple differential equation, and the solution for the central deflection d of the strut is

$$d = e \sec \frac{l}{2}\sqrt{\frac{W}{EI}} \qquad (1.5)$$

This relation is plotted in Fig. 1.2. It will be seen that it is highly non-linear, and indeed d tends to infinity as the load W tends to the Euler buckling load $W_E = \pi^2 EI/l^2$. However, if the

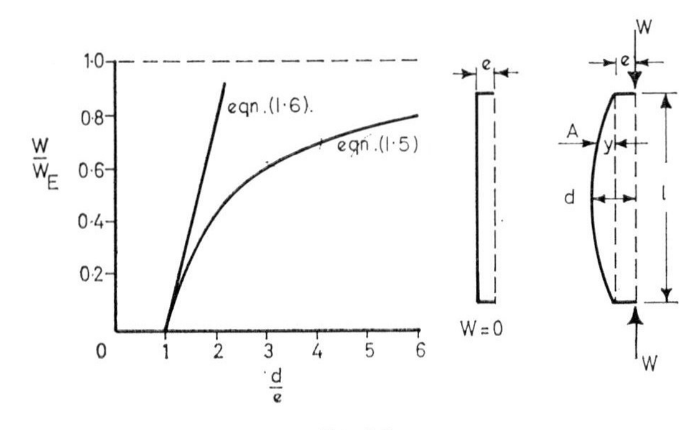

FIG 1.2

influence of deflections on the equilibrium equation was ignored, equation (1.4) would be replaced by $We = M$. The corresponding relation between W and d would then be the linear relation

$$d = e\left[1+\frac{Wl^2}{8EI}\right] \qquad (1.6)$$

This relation is also shown in Fig. 1.2. It will be seen that the discrepancy between equations (1.5) and (1.6) becomes large as W increases.

1.4. STATICALLY DETERMINATE AND INDETERMINATE STRUCTURES

In some structures, including many of considerable practical importance, it is possible to find the distribution of all the internal forces by considering only the requirements of statical

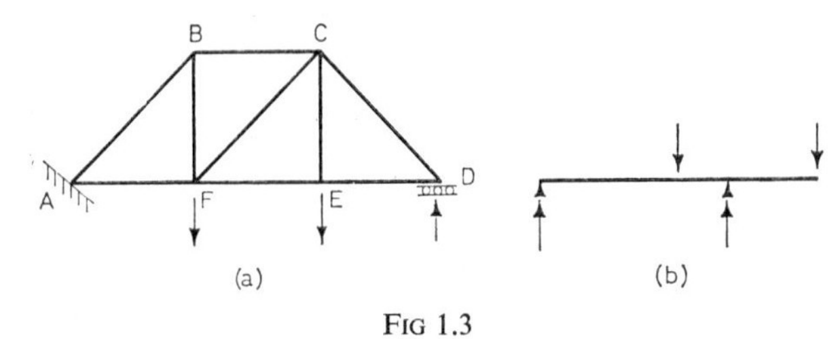

FIG 1.3

equilibrium. Some simple examples are shown in Fig. 1.3. For such structures the complete structural analysis assumes an especially simple form. The internal forces are first found by considering statical equilibrium. Then from these forces, the deformations of the members are determined from the appropriate member characteristics. There then remains the purely

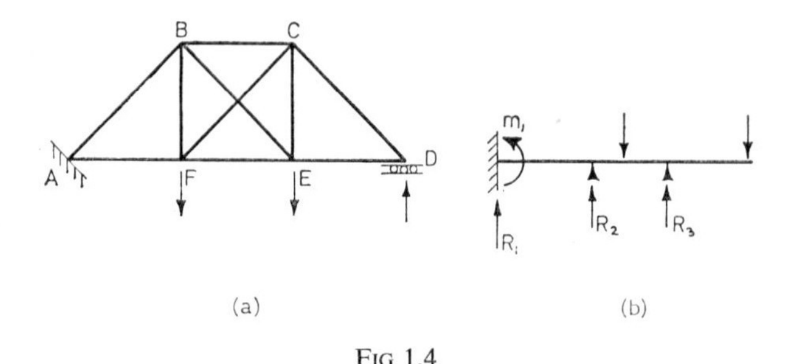

FIG 1.4

geometrical problem of determining the deflected form of the structure as a whole from a knowledge of the deformations of the individual members.

When it is not possible to determine the distribution of all the internal forces by statics alone, the structure is said to be statically indeterminate. Some examples of statically indeterminate structures are shown in Fig. 1.4. The truss of Fig. 1.4(a) can be compared with the truss of Fig. 1.3(a), and it will be seen that an additional bar BE has been inserted. Since the truss of Fig. 1.3(a) is statically determinate, it is evident that the forces in all the bars of the truss of Fig. 1.4(a) are known if the force R in the member BE is specified. Thus for example the force P in the member BF may be written as

$$P = P_W + rR \tag{1.7}$$

where P_W is the force in BF if the force in BE is zero, and r is a numerical coefficient.

Viewed from this standpoint, the bar BE is redundant in the sense that it is not necessary for the formation of a statically determinate structure. The truss of Fig. 1.4(a) is said to have one redundancy, and equation (1.7) expresses the fact that the force in a typical bar can be written in terms of known quantities and the force R in the redundant bar.

It will of course be realised that the redundant bar could equally be chosen as any one of the six bars BC, CE, EF, BF, BE, or CF in the central panel. The other four bars AB, AF, DC, and DE could not be selected as the forces in these bars can be determined by statics alone.

The beam of Fig. 1.4(b) has two redundancies, since if the two props with reactions R_2 and R_3 were removed a statically determinate cantilever would remain. If the bending moment at any section due to the loading acting upon this cantilever is denoted by M_W, the bending moment M at any section could be expressed as

$$M = M_W + r_2 R_2 + r_3 R_3, \tag{1.8}$$

where r_2 and r_3 are coefficients which are functions of position along the beam. It would, however, be equally valid to regard m_1 and R_1 as the two redundancies, in which case M could be expressed as

$$M = M'_W + \alpha_1 m_1 + r_1 R_1 \tag{1.9}$$

where again α_1 and r_1 are functions of position along the beam. In fact, M can be expressed in terms of any pair of the four quantities R_1, R_2, R_3 and m_1.

For redundant structures the complete structural analysis is necessarily more complex than for statically determinate systems. The three basic conditions of equilibrium, compatibility, and the member characteristics must still be expressed, but all of these conditions will need to be taken into account simultaneously to obtain a solution to the problem. For instance, in certain procedures the conditions of equilibrium are first used to express each internal force in terms of the redundancies, as for example in equation (1.8). The values of the redundancies are then determined by considerations of compatibility. This involves using the member characteristics to express the deformations in terms of the redundancies, and the correct values of the redundancies are then those which enable all the requirements of compatibility to be satisfied. As will be seen in the next section, another approach is possible, but again it is necessary to consider the three basic conditions together in determining a solution.

1.5. CHOICE OF VARIABLES

Two fundamentally different approaches may be adopted for the analysis of redundant structures (Charlton, 1956). In one of these, termed the compatibility method, attention is first directed towards determining the correct values of the redundancies which enable all the requirements of compatibility to be met. The remaining internal forces and the deflections are then found. In the other approach, termed the equilibrium method, the initial step is the determination of the correct values of the deflections of the structure so that all the requirements of equilibrium are met. The internal forces in the members are then deduced.

These two approaches will be illustrated with reference to the symmetrical truss illustrated in Fig. 1.5. In this truss the three bars are all of the same cross-section and material, with area A and Young's Modulus E, and the loading is such that each bar behaves elastically. It is required to find the forces in the bars and the vertical deflection of F due to the vertical load W.

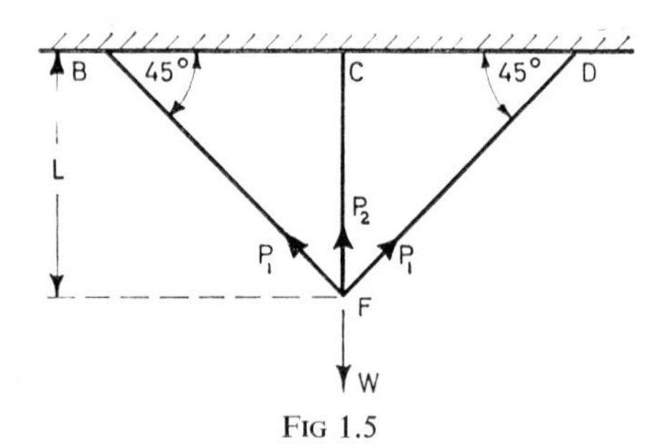

FIG 1.5

Compatibility method

In this method consideration is first given to all the requirements of equilibrium; in this case there is simply one equation of vertical equilibrium of F, namely

$$W = P_2+\sqrt{2}\,.\,P_1, \tag{1.10}$$

horizontal equilibrium of F having been ensured by making the tensions P_1 in BF and DF equal because of symmetry.

Either P_1 or P_2 can be chosen as the single redundancy of the system. Choosing P_1 arbitrarily, it follows from equation (1.10) that

$$P_2 = W-\sqrt{2}\,.\,P_1, \tag{1.11}$$

and the primary objective is the determination of the redundancy P_1.

If e_1 and e_2 are the extensions of the bars BF (or DF) and CF, respectively, and Δ is the vertical deflection of F, the requirements of geometrical compatibility are

$$\Delta = \sqrt{2}\,.\,e_1 = e_2 \tag{1.12}$$

The member characteristics are those of elasticity, and hence

$$e_1 = \frac{P_1\sqrt{2}\,.\,L}{AE} \tag{1.13}$$

$$e_2 = \frac{P_2 L}{AE} \tag{1.14}$$

P_2 has already been expressed in terms of P_1 by equation (1.11), and substituting in equation (1.14) it follows that

$$e_2 = \frac{(W-\sqrt{2}\,.\,P_1)L}{AE}$$

e_1 and e_2 are now substituted in equation (1.12) to give a single compatibility equation for the determination of P_1, namely

$$\sqrt{2}\left[\frac{P_1\sqrt{2}\,.\,L}{AE}\right] = \frac{(W-\sqrt{2}\,.\,P_1)L}{AE},$$

from which $$P_1 = W\left(1-\frac{1}{\sqrt{2}}\right)$$

To complete the solution, P_2 is found from equation (1.11) to be $W(2-\sqrt{2})$, and Δ is found from equations (1.12) and (1.13) to be $WL(2-\sqrt{2})/AE$.

Equilibrium method

In this method the primary variable which is sought is Δ, the vertical deflection of joint F, and attention is first directed towards the requirements of compatibility. These were expressed above as equation (1.12), namely

$$\Delta = \sqrt{2}\,.\,e_1 = e_2$$

This equation gives the bar elongations e_1 and e_2 in terms of Δ. The next step is to use the member characteristics to express

P_1 and P_2 in terms of Δ. P_1 and P_2 were given in terms of e_1 and e_2 by equations (1.13) and (1.14), and it follows that

$$P_1 = \frac{AE}{\sqrt{2}.L} e_1 = \frac{AE}{2L} \Delta \tag{1.15}$$

$$P_2 = \frac{AE}{L} e_2 = \frac{AE}{L} \Delta \tag{1.16}$$

The appropriate value of Δ which enables the condition of equilibrium, equation (1.10), to be satisfied, is now determined by substituting the above values of P_1 and P_2 in this equation, giving

$$W = \frac{AE}{L} \Delta + \sqrt{2}.\left(\frac{AE}{2L} \Delta\right),$$

whence
$$\Delta = \frac{WL}{AE} (2 - \sqrt{2})$$

The solution is then completed by determining P_1 and P_2 from equations (1.15) and (1.16), giving the same results as before.

For the simple example of Fig. 1.5 there was little to choose between the equilibrium and compatibility approaches. In each case there was one primary variable to be determined, either P_1 or Δ. However, in more complex structures there may be a considerable advantage in one method or the other. Consider for instance the pin-jointed truss illustrated in Fig. 1.6(a). This structure has only one redundancy, so that if the compatibility method was used there would be one primary variable to be determined. However, there are 17 unknown joint displacements, namely two each at joints B, C, D, E, G, H, I and J, and one at F, and these would need to be determined from 17 simultaneous equations of equilibrium if the equilibrium method was used. Clearly the compatibility method is far superior in such a case.

For the frame illustrated in Fig. 1.6(b) there are 18 redundancies which would need to be determined if the compatibility method was used. However, if the equilibrium method was used there would be 12 deformation variables to be found,

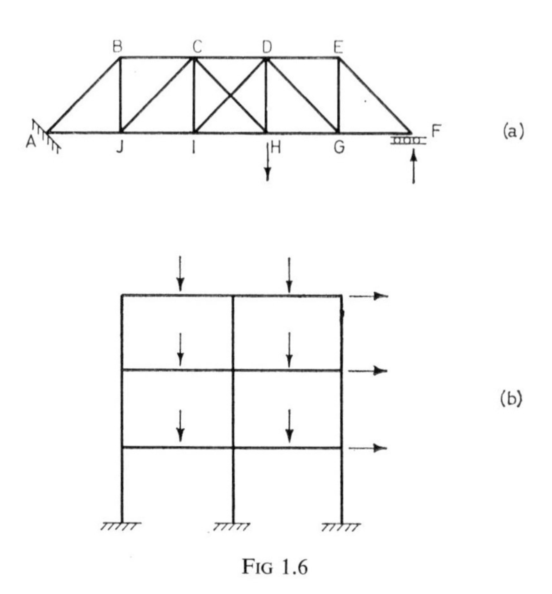

FIG 1.6

these being the 9 unknown joint rotations and the sways of each of the three storeys. In this case the equilibrium method suggests itself as better adapted to the problem.

1.6. DIRECT AND INDIRECT APPROACHES

In the compatibility and equilibrium approaches to the problem of Fig. 1.5 which were given in the preceding section, the conditions of equilibrium and compatibility were written down directly. This presented no particular difficulty; the equilibrium equation was obtained immediately by vertical resolution at joint F, and the compatibility equation was also derived directly from a simple consideration of the geometry of a small vertical displacement of F. However, in more complex problems it may prove to be very tedious, although not difficult in principle, to establish the requisite conditions of equilibrium or of compatibility, and recourse is then often had to indirect methods for deriving these conditions.

It is the primary purpose of this book to describe the ideas underlying these indirect methods and to show that they can all be developed from the Principle of Virtual Work. When applied to specific types of problem, this principle yields methods of analysis which are precisely equivalent to those which are derived by correct applications of the appropriate energy methods. The relationships between the Principle of Virtual Work and these other principles and theorems are therefore carefully emphasised in the book.

If the compatibility approach is adopted for the solution of a particular problem, the compatibility conditions need not be derived directly. Instead, the actual geometrical problem of determining these conditions may be transformed by an application of the virtual work principle into a hypothetical equilibrium problem, the solution of which enables the compatibility conditions to be written down. This transformation would, of course, only be used if the hypothetical equilibrium problem was easier to solve than the actual geometrical problem.

Alternatively, if the equilibrium approach is adopted, the equilibrium conditions need not necessarily be derived directly by statics. By an application of virtual work it is found that the solution of a hypothetical geometrical problem can be used to establish the equilibrium conditions which are required. Again, this transformation would only be used if the hypothetical geometrical problem was easier to solve than the actual equilibrium problem.

2

Principles of superposition

2.1. INTRODUCTION

As the term implies, superposition as a general principle is concerned with situations in which two or more influences act together. If their separate effects are known, superposition asserts that these effects are additive, so that for example

if Cause A produces Effect A
and Cause B produces Effect B
then Cause A + Cause B produces Effect A + Effect B

Although the point is not often made clear, there are several distinct Principles of Superposition. The one most often quoted refers to linear elastic structures; this particular principle is described in Section 2.4, and this is preceded by discussions of the conditions under which force systems and displacement systems can be superposed. The chapter concludes by considering applications of superposition which make use of either symmetry or skew-symmetry.

2.2. SUPERPOSITION OF FORCE SYSTEMS

In discussing the superposition of force systems it is necessary to distinguish between the cases of statically determinate and indeterminate structures. Consideration will first be given to statically determinate structures.

Statically determinate structures

For statically determinate structures the Principle of Superposition of forces has quite general validity subject only to one proviso, namely that the distortions of the structure must not be gross in the sense defined in Section 1.3, so that the equations of equilibrium are sensibly the same as for the undistorted structure.

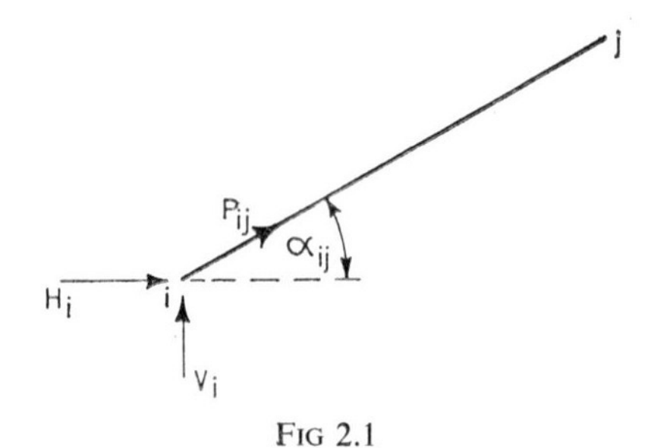

FIG 2.1

If this condition is fulfilled, the internal forces and moments in any statically determinate structure due to two different loading systems acting together may be found by adding together the internal forces and moments caused by each of the loading systems acting separately.

The precise form taken by the proof of this principle depends upon the particular type of structure under consideration. For a pin-jointed truss the proof is as follows:

Consider any typical joint i in the truss, and let ij be a typical bar radiating from this joint, the tension in this bar being denoted by P_{ij} as in Fig. 2.1. It is supposed that the external load at this joint has horizontal and vertical components H_i and V_i, as shown; these could be either applied loads or else support reactions.

As already stated, the distortion of the truss under load can be ignored in establishing the equations of equilibrium, so that for example the angle α_{ij} which the bar ij makes with the horizontal is the same as for the unloaded truss. The equations of equilibrium for the joint i are thus

$$\left.\begin{aligned} H_i + \sum_j P_{ij} \cos \alpha_{ij} &= 0 \\ V_i + \sum_j P_{ij} \sin \alpha_{ij} &= 0 \end{aligned}\right\} \tag{2.1}$$

where the angles α_{ij} do not depend on the loads. The summations in equations (2.1) cover all the bars radiating from the joint i.

Similar pairs of equations can be written for every joint in the truss, and if the truss is statically determinate there will be just a sufficient number of equations to determine the forces P_{ij} in all the bars, together with any unknown reactions due to the prescribed loading.

The Principle of Superposition for forces can now be stated as follows: "If the set of external loads (H_i', V_i') causes forces P_{ij}' in the bars of a statically determinate truss, and the set of external loads (H_i'', V_i'') causes forces P_{ij}'', then the set of external loads ($H_i'+H_i''$, $V_i'+V_i''$) will cause forces $P_{ij}'+P_{ij}''$."

To establish this result, it is noted that the two sets of equilibrium equations are as follows:

$$\left.\begin{aligned} H_i' + \sum_j P_{ij}' \cos \alpha_{ij} &= 0 \\ V_i' + \sum_j P_{ij}' \sin \alpha_{ij} &= 0 \end{aligned}\right\} \tag{2.2}$$

$$\left.\begin{aligned} H_i'' + \sum_j P_{ij}'' \cos \alpha_{ij} &= 0 \\ V_i'' + \sum_j P_{ij}'' \sin \alpha_{ij} &= 0 \end{aligned}\right\} \tag{2.3}$$

Adding these sets of equations, it is found that

$$\left.\begin{aligned} (H_i'+H_i'') + \sum_j (P_{ij}'+P_{ij}'') \cos \alpha_{ij} &= 0 \\ (V_i'+V_i'') + \sum_j (P_{ij}'+P_{ij}'') \sin \alpha_{ij} &= 0 \end{aligned}\right\} \tag{2.4}$$

Comparing the sets of equations (2.4) and (2.1) it is seen that equations (2.4) represent the conditions of equilibrium between the bar forces $P'_{ij}+P''_{ij}$ and the set of external loads $(H'_i+H''_i, V'_i+V''_i)$. Because the system under discussion is statically determinate, equations (2.4) must have a unique solution, so that the principle has been established.

An obvious corollary of this result is that if a set of loads (H_i, V_i) causes bar forces P_{ij}, then the set of loads (kH_i, kV_i), where k is a numerical coefficient, will produce bar forces kP_{ij}.

It will be seen from the proof that the principle depends solely upon the fact that the equations of equilibrium (2.1) are linear in form because of the assumed constancy of the angles α_{ij}. Once these linear equations are established, and since there is just the right number of equations for the unique determination of the bar forces, the result follows immediately. Consequently, it is a simple matter to establish the principle for other types of statically determinate systems, notably beams, and further proofs will therefore not be set out in formal terms.

Statically indeterminate structures

It is the characteristic of statically indeterminate or redundant structures that the set of equilibrium equations such as equations (2.1) for a truss system does not provide enough information for the determination of the internal forces. Consideration of superposition for redundant structures in general will be deferred until Section 2.4. However, there is a special sense in which the result just derived can be applied to indeterminate structures.

Consider for example the truss shown in Fig. 2.2(a). This would usually be analysed by the compatibility method, and it will be supposed that the forces R_1 and R_2 in the bars CF and BG, respectively, are selected as the two redundancies which are sought as the primary variables. A convenient way of setting out the statical analysis is to consider in turn the bar forces for the three situations depicted in Figs. 2.2(b), (c) and (d).

Figure 2.2(b) shows the truss with the two redundant bars removed. The statically determinate truss so formed is termed the basic truss. The loading is as for Fig. 2.2(a). Figure 2.2(c) shows the basic truss with no external load but subjected to the unit forces at C and F which would result from a unit tension in CF, and Fig. 2.2(d) shows the basic truss similarly loaded by unit tension in BG.

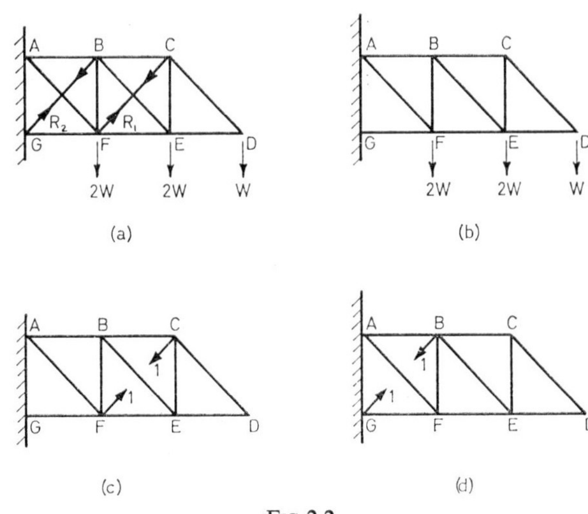

FIG 2.2

The Principle of Superposition for forces can be applied to the basic truss, which is statically determinate. Suppose that the force in a typical bar in Fig. 2.2(b) is P_W, and for Figs. 2.2(c) and (d) let the forces in the same bar be r_1 and r_2, respectively. Then it follows from superposition that the actual tension P in this typical bar when the loading is applied and the redundancies have the values R_1 and R_2 will be given by

$$P = P_W + r_1 R_1 + r_2 R_2 \tag{2.5}$$

As will be seen in Chapter 4, it is convenient in analyses by the compatibility method to tabulate the forces P_W, r_1 and r_2

for each bar. After the redundancies have been determined from the compatibility equations, the actual bar forces are then found from equations such as equation (2.5).

It should be pointed out that equation (2.5) can be applied to the two redundant bars as well as to the bars of the basic truss. For each redundant bar, $P_W = 0$; for CF, $r_1 = 1$ and $r_2 = 0$, whereas for BG, $r_1 = 0$ and $r_2 = 1$.

The superposition of forces as exemplified in equation (2.5) holds true irrespective of whether or not Hooke's Law is obeyed by the members. The linearity required for superposition

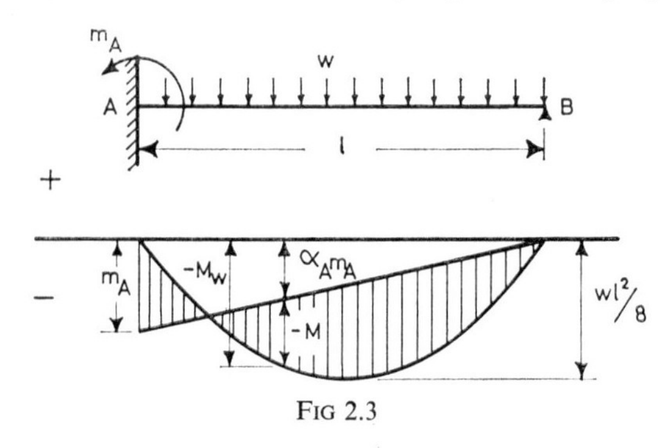

FIG 2.3

is ensured by the assumption of deflections which are sufficiently small to enable the equations of equilibrium to be written down for the undistorted structure.

Another example of the application of this principle is afforded by the propped cantilever illustrated in Fig. 2.3. The single redundancy for this beam may be taken as the hogging clamping moment m_A. If this moment was zero, the bending moment at any section would be the "free" bending moment, denoted by M_W, which varies parabolically as indicated in the figure with a maximum value of $-wl^2/8$. (A negative sign is used here because hogging moments are taken as positive). If the load was absent, but m_A was applied, the bending moment at any section would be the "reactant" moment, denoted by $\alpha_A m_A$, where α_A is a linear function of distance along the beam which is unity at A and zero at B. By the Principle of Superposition, the actual bending moment M at any section would then be

$$M = M_W + \alpha_A m_A \tag{2.6}$$

Thus if the reactant moment is plotted with its sign reversed, as in the figure, the actual bending moment is the difference in ordinate between the free and reactant lines. This construction is shown in Fig. 2.3, from which it is seen that $-M = -M_W -\alpha_A m_A$, in accordance with equation (2.6).

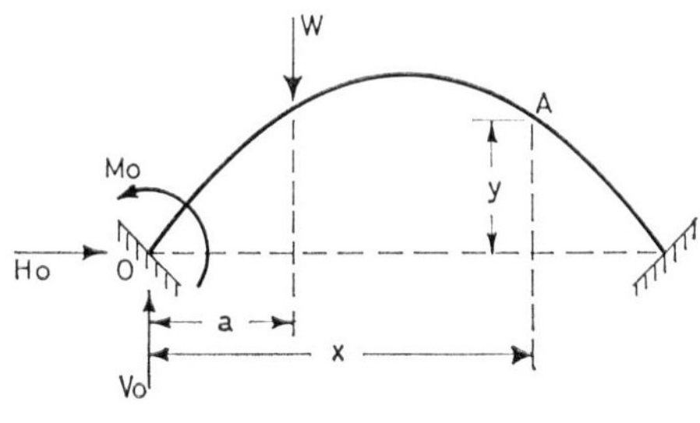

FIG 2.4

A final example of the application of this principle is illustrated in Fig. 2.4. The arch has three redundancies, which might be selected as the horizontal and vertical components of reaction H_0 and V_0 at the left-hand abutment, together with the clamping moment M_0. The hogging bending moment M at section A has the value

$$M = M_0 + H_0 y - V_0 x + W(x-a) \tag{2.7}$$

This is justified by imagining the arch to be completely free at the left-hand abutment, thus rendering it statically determinate. The separate effects of M_0, H_0, V_0 and W are then superposed.

2.3. SUPERPOSITION OF DISPLACEMENTS

A Principle of Superposition also holds true for displacements provided that the deflections of the structure are not gross.

This principle is true for both statically determinate and indeterminate structures. A proof will now be given for trusses.

Figure 2.5 shows a typical bar ij of length l_{ij} in a pin-jointed truss. This bar, which makes an angle α_{ij} with the horizontal, is supposed to undergo an extension δl_{ij} due to any cause whatsoever.† In addition, the ends i and j are supposed to undergo displacements whose horizontal and vertical components are (h_i, v_i) and (h_j, v_j), respectively.

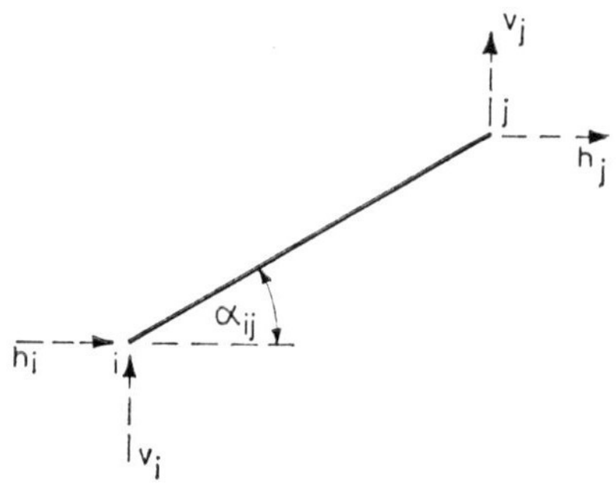

FIG 2.5

For geometrical compatibility the extension δl_{ij} of this bar must equal the difference in the axial movement at the two ends, so that

$$\begin{aligned} \delta l_{ij} &= h_j \cos \alpha_{ij} + v_j \sin \alpha_{ij} - h_i \cos \alpha_{ij} - v_i \sin \alpha_{ij} \\ &= (h_j - h_i) \cos \alpha_{ij} + (v_j - v_i) \sin \alpha_{ij} \qquad (2.8) \end{aligned}$$

An equation of this kind can be written down for every bar of the truss, and these equations must be satisfied by any compatible set of joint displacements and bar elongations. Because of the assumption that α_{ij} is sensibly the same as for the undistorted truss, these equations are all linear. It therefore

† δl is used throughout the book to denote the total extension of a bar due to all causes (e.g., force, temperature, etc.). The symbol e is reserved for that part of the extension which is caused by the axial force in the bar.

follows at once that if a set of bar elongations $(\delta l)'_{ij}$ is compatible with a set of joint displacements (h'_i, v'_i), and a set of bar elongations $(\delta l)''_{ij}$ is compatible with a set of joint displacements (h''_i, v''_i), then the set of bar elongations $(\delta l)'_{ij}+(\delta l)''_{ij}$ will be compatible with the set of joint displacements $(h'_i+h''_i, v'_i+v''_i)$. This is the Principle of Superposition for displacements.

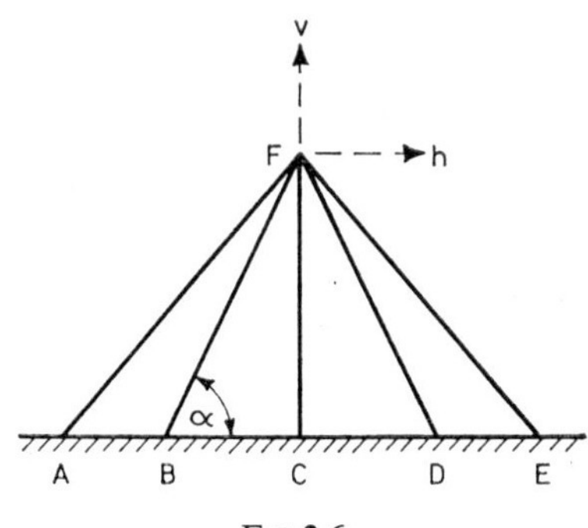

FIG 2.6

The principle is widely used in connection with the equilibrium method of analysis. Consider for example the truss shown in Fig. 2.6. The primary objective of the equilibrium method of analysis for this structure would be to determine the horizontal and vertical components of deflection, h and v, of the joint F. The initial step would therefore be to express the elongation of each bar in terms of h and v. This may be done by considering two separate deformation systems. In the first of these, there is a (small) unit horizontal displacement of F, with v zero, and in the second there is a (small) unit vertical displacement of F, with h zero. The corresponding elongations of the typical bar BF in these two systems are denoted by $(\delta l)_h$ and $(\delta l)_v$, respectively, and it is seen at once that

$$(\delta l)_h = \cos\alpha$$

$$(\delta l)_v = \sin\alpha$$

From the Principle of Superposition of displacements it follows that if δl is the elongation of the typical bar BF when the displacement components at F are h and v,

$$\begin{aligned} \delta l &= h(\delta l)_h + v(\delta l)_v \\ &= h \cos \alpha + v \sin \alpha \end{aligned} \tag{2.9}$$

This relation holds true whatever the cause of the displacements, which might be due to load, temperature change, lack of fit, etc., and is also true whether or not Hooke's Law is obeyed by the members. It could of course have been derived without recourse to the Principle of Superposition; however, it will be seen in Chapter 6 that displacement systems such as $(\delta l)_h$ and $(\delta l)_v$ are required in the equilibrium method of analysis.

Most other examples of the application of this principle are implicit in the examples to be discussed in the next section, which deals with the Principle of Superposition for structures whose members obey Hooke's Law.

2.4. SUPERPOSITION FOR LINEAR ELASTIC STRUCTURES

If the members of a structure all obey Hooke's Law, and in addition the deformations are not gross, the structure is referred to as linear elastic. For such structures, a general Principle of Superposition can readily be established. According to this principle, the changes in the deflections and the internal forces of a linear elastic structure due to applied loading, temperature change, movement of supports, etc., can all be superposed.

This principle will first be established for initially stress-free pin-jointed trusses whose members obey Hooke's Law, and in which the deformations are solely due to external loads. If the tension in a typical bar ij is P_{ij}, and the extension of this bar is e_{ij}, then Hooke's Law states that

$$P_{ij} = \mu_{ij} e_{ij} \tag{2.10}$$

where μ_{ij} is the elastic constant for the bar. If the bar is uniform and of original length l_0 and cross-sectional area A_0, then $\mu = EA_0/l_0$, where E = Young's Modulus. Substituting this relation in the compatibility equation (2.8) it is seen that

$$\frac{P_{ij}}{\mu_{ij}} = (h_j - h_i)\cos\alpha_{ij} + (v_j - v_i)\sin\alpha_{ij} \qquad (2.11)$$

In Section 2.2 it was shown that the two equations of equilibrium for a typical joint i were as follows:

$$\left.\begin{aligned} H_i + \sum_j P_{ij}\cos\alpha_{ij} &= 0 \\ V_i + \sum_j P_{ij}\sin\alpha_{ij} &= 0 \end{aligned}\right\} \qquad (2.1)$$

Equations (2.11) and (2.1), expressing the conditions of equilibrium and compatibility, together with the member characteristics, must be sufficient for a complete solution of the problem. This can be seen from the following considerations. If the structure has b bars and j joints, there will be b unknown bar forces such as P_{ij}. Of the $2j$ displacement components such as (h_i, v_i), suppose that r are prescribed at the supports, leaving $(2j-r)$ unknown displacements. Of the external forces (H_i, V_i), some will be prescribed external loads, but there will be r unknown reactions, one corresponding to each support restraint. The total number of unknowns is therefore as follows:

Bar forces	b
Displacements	$2j-r$
Reactions	r
Total	$b+2j$

There will be an equation of compatibility such as equation (2.11) for each bar, so that there will be b equations of compatibility. Further, there will be $2j$ equations of equilibrium, equations (2.1) representing a typical pair for a joint. The total number of equations of equilibrium and compatibility, the latter also incorporating the member characteristics, will therefore be $b+2j$, so that there is just the right number of equations for the determination of all the unknowns. Each of these equations is linear in form, and the Principle of Superposition follows immediately.

The proof is readily extended to cases in which the deformations are not due solely to external loads. Consider for example the question of extensions due to temperature change. The stress/strain law of equation (2.10) is then replaced by the member characteristic

$$\delta l_{ij} = \frac{P_{ij}}{\mu_{ij}}\beta + {}_{ij}l_{ij}T_{ij} \tag{2.12}$$

where β_{ij} is the linear coefficient of expansion† for the bar ij, l_{ij} is its length, and T_{ij} its temperature rise above some datum. The equations (2.1) of equilibrium remain unaffected, and equation (2.11) is replaced by

$$\frac{P_{ij}}{\mu_{ij}} + \beta_{ij}l_{ij}T_{ij} = (h_j - h_i)\cos\alpha_{ij} + (v_j - v_i)\sin\alpha_{ij} \tag{2.13}$$

Equations (2.13) still constitute b linear equations, and so by the same argument as before the Principle of Superposition still holds.

Further extensions to cover such cases as the movement of supports or the lack of fit of members are also easily made, and will not be given here. Nor is it necessary to set out formal proofs for other types of structure. It should now be obvious that the Principle of Superposition can be applied to any situation in which the complete structural analysis is linear in character. The two necessary conditions for the general Principle of Superposition to hold true are therefore as follows:

† The symbol β is used here to denote the linear coefficient of expansion to avoid confusion with α, which denotes the angle between the direction of a typical bar and the horizontal.

(i) The deformations must not be gross, so that the conditions of equilibrium and compatibility can be written down without loss of accuracy using the original orientations of the members.

(ii) There must be no non-linearity in the member characteristics.

It should perhaps be emphasised that the principle deals with changes in deformations and internal forces. Care must therefore be taken in defining the datum from which the deformations and internal forces are measured. Thus if a structure is initially free from stress when unloaded, its configuration in this stress-free condition would be the natural datum to consider. Suppose that due to sinking of supports, a component of deflection at a certain joint is d_0, and that subsequently two load systems A and B may be applied separately or together. If d_A and d_B are respectively the deflections caused by these two load systems, the total deflections at the joint when the loads are applied would be as follows:

$$\begin{aligned} \mathrm{A} &: d_0+d_A \\ \mathrm{B} &: d_0+d_B \\ \mathrm{A+B} &: d_0+d_A+d_B \end{aligned}$$

Thus as far as the loads are concerned, it is the changes in deflection which may be superposed, rather than the total deflections measured from the stress-free configuration. A similar argument also applies to the internal forces.

The importance and utility of this principle for linear elastic structures will be immediately obvious. Thus for any such structure, whether statically determinate or indeterminate, the effects produced by two different loading systems can be superposed. A building frame may be subjected to wind and snow loads in varying proportions, but their separate effects need only to be calculated and then superposed. Again, if in a loaded structure a support sinks, the separate effects of the sinking can be calculated and then superposed on the effects due to the loading.

Apart from such obvious uses of the principle, there are often analytical advantages in splitting the loading on a symmetrical frame into its symmetrical and skew-symmetrical components. This question will be considered in detail in Section 2.5.

The elastic behaviour of members is not a sufficient condition for the principle to be applicable. Consider for example the eccentrically loaded elastic strut of Fig. 1.2. The central deflections produced by loads W_1 and W_2 are given by equation (1.5) as

$$d_1 = e \sec \frac{l}{2}\sqrt{\frac{W_1}{EI}}$$

$$d_2 = e \sec \frac{l}{2}\sqrt{\frac{W_2}{EI}}$$

and the deflection d due to a load (W_1+W_2) is

$$d = e \sec \frac{l}{2}\sqrt{\frac{(W_1+W_2)}{EI}}$$

Clearly d is not the sum of d_1 and d_2, and this is also obvious from a study of the (W, d) relation of Fig. 1.2. The reason for the failure of the Principle of Superposition in this case is that the deformations are gross, in the sense that the conditions of equilibrium are appreciably affected by the distortion of the structure.

2.5. SYMMETRY AND SKEW-SYMMETRY

The approach to many problems can be simplified if the structure is symmetrical and in addition the loading or other deformation producing agency is either symmetrical or skew-symmetrical (also known as anti-symmetrical). The former case will be considered first.

Symmetrical structure and loading

Several examples of symmetrical structures are shown in Fig. 2.7. In each case the structure is completely symmetrical; thus in Fig. 2.7(a) the shape of the portal frame is symmetrical about a vertical axis through C, and in addition the members AB and ED are identical, as are the members BC and DC. As

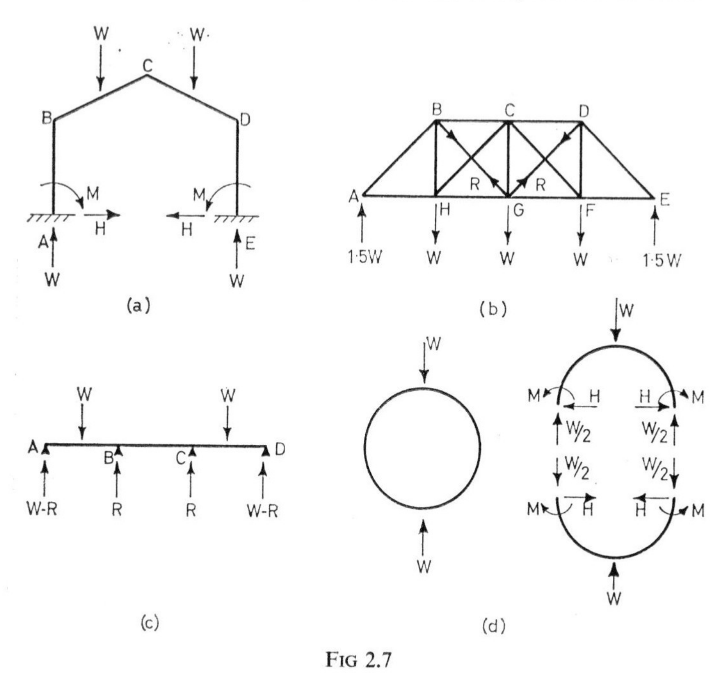

FIG 2.7

will be seen, the loading is also symmetrical in each case, and it will also be supposed that any other agencies which produce internal forces and deformations also act symmetrically. Thus, for example, in Fig. 2.7(a) it is presumed that if the support A was rotated through an angle θ counter-clockwise, then the support E would rotate by θ clockwise. Other examples of symmetrical actions for this structure would be a spreading apart of the feet A and E or a uniform rise in temperature of all the members.

It will be supposed that each of these structures would be analysed by the compatibility method, so that the starting point of each analysis would be the specification of the redundancies. It will be shown that because of symmetry the number of redundancies in each case is less than for a general loading case.

Consider first the portal frame of Fig. 2.7(a). Because of symmetry, the vertical reactions at A and E must be equal, and are therefore both W, as shown. Furthermore, the horizontal reactions H and the restraining moments M at A and E must be equal and opposite. It will then be seen that H and M may be taken as the two redundancies for this problem. In the absence of symmetry there would, of course, be three redundancies, so that by making use of symmetry the analysis is simplified.

The argument may be made more formal if desired by noting that if the frame and loading are viewed from the rear, with A and E, and also B and D interchanged, the frame and its loading are unchanged. It follows at once that the reactions at A and E are equal, so that by vertical resolution each is of magnitude W. Further, the horizontal reactions and moments must also be equal.

A point of particular importance is that the above arguments are not dependent in any way upon the assumption of elastic behaviour. Hence the conclusions reached are equally valid whether or not Hooke's Law is obeyed; they would, for instance, still be true if the frame was at its plastic collapse load, with a sufficient number of plastic hinges present to reduce the structure to a mechanism.

For the symmetrical truss of Fig. 2.7(b) there would be two redundancies for a general non-symmetrical loading. However, if the truss was initially free from stress, and was then subjected to the symmetrical loading shown, the forces in the members BG and DG would be equal, and the force R in each of these members could be taken as the single redundancy for this case.

The beam of Fig. 2.7(c) would in general have different reactions at B and C, and these would constitute the two redundancies of the problem. If the beam was initially free from stress, and was then loaded symmetrically as shown, there would be equal reactions R at B and C, and the reactions at A and D would also be equal. These latter reactions would each be $(W-R)$, by vertical resolution, and so there would only be a single redundancy R. A similar argument would also hold true in this case if the supports B and C sank through equal distances when the symmetrical load was applied.

Finally, the proving ring of Fig. 2.7(d), if subjected to a general loading in its own plane, would have three redundancies. However, a consideration of the top half of the ring shown, with symmetrical loading, leads to the conclusion that the internal forces and moments must be as indicated, the reasoning being the same as for the portal frame of Fig. 2.7(a). The equal and opposite reactions and moments on the bottom half of the ring are also shown. It will be seen that the shear force H is tending to reduce the diameter of the lower portion, whereas for the upper portion H tends to increase the diameter. Since the two halves are experiencing similar loading conditions, H must therefore be zero. It will be noticed that for the two halves of the ring, M acts in each case so as to decrease the radius of curvature, and so no conclusion can be drawn about the value of this bending moment. M is thus left as the single redundancy of the problem.

Symmetrical structure and skew-symmetrical loading

Useful deductions can often be made when a symmetrical structure is subjected to skew-symmetrical loading. Consider for example the symmetrical pin-jointed arch shown in Fig. 2.8(a), which will be supposed to be free from stress when unloaded. Under a general loading this arch would have one redundancy. For the skew-symmetrical loading shown, suppose that the horizontal reactions at the abutments are each H, the vertical reactions being determined by statics as shown.

Now consider the view of this arch from the rear, as in Fig. 2.8(b). It will be seen that the loads are now changed in sign, but the structure itself is unaltered because of its symmetry. Figures 2.8(a) and (b) may therefore be regarded as depicting two different loadings on the same structure.

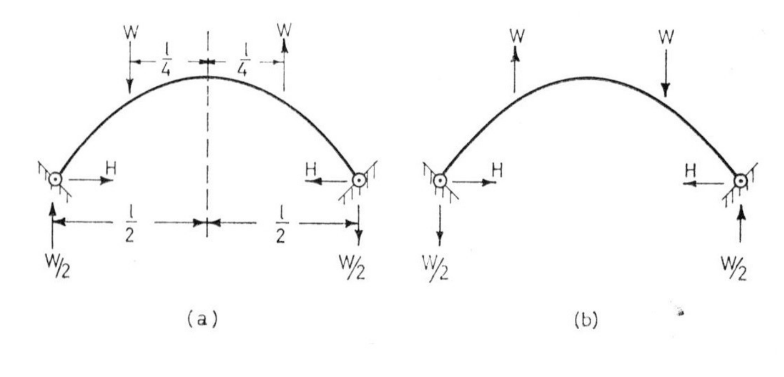

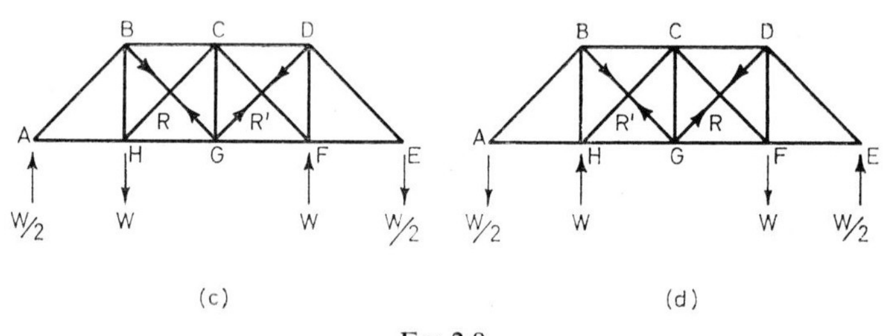

FIG 2.8

If Hooke's Law is obeyed, so that the structure is linear elastic, the effects of these two loadings may be superposed. It follows that the two loadings superposed, giving zero applied load, will produce an abutment thrust $2H$, and so H must in fact be zero. The problem has therefore become statically determinate.

A similar argument could have been used instead to show

that the bending moment at the crown of the arch must be zero; this moment might have been selected as the redundancy instead of H. It is easily seen that these two results are equivalent. Further, it can be shown that the vertical deflection of the crown of the arch must be zero.

If Hooke's Law is not obeyed, these results will still be true provided that the bending moment–curvature relation for the arch member is symmetrical, so that sagging and hogging bending moments of the same magnitude produce sagging and hogging curvatures which are also of the same magnitude. In this case it is evident that a complete reversal of load would be expected to change the signs of all the reactions and internal forces and moments without affecting their magnitudes. Comparison of Figs. 2.8(a) and (b) then shows that H would be zero. The deflections would also be of the same magnitude but with the signs changed, and so the vertical deflection at the crown of the arch would also be zero.

A further illustration of skew-symmetry is afforded by the linear elastic truss of Fig. 2.8(c), which is symmetrical and free from stress when unloaded. In this truss the two redundant bars are selected as GB and GD, with forces R and R', respectively. Figure 2.8(d) shows the same structure and loading when viewed from the rear. By superposing the effects of these two loadings, a situation results in which there is no load on the structure but the forces in GB and GD are each $(R+R')$. It follows that $R' = -R$, thus reducing the problem to a singly redundant situation.

Alternatively, it could be shown that the force in the bar GC is zero; this however may be deduced from the fact that $R = R'$ by considering vertical equilibrium of the joint G, and so does not give rise to a further reduction in the degree of redundancy.

Symmetrical and skew-symmetrical components of load

Any loading on a symmetrical structure may be regarded as the sum of two loadings, one of which is symmetrical and the other skew-symmetrical. Thus for example the loading on the arch shown in Fig. 2.9(a) can be regarded as the sum of the loadings of Figs. 2.9(b) and (c). Provided that the arch is linear elastic, so that the Principle of Superposition can be applied, the effects of the loading of Fig. 2.9(a) will be the sum of the effects of the loadings of Figs. 2.9(b) and (c).

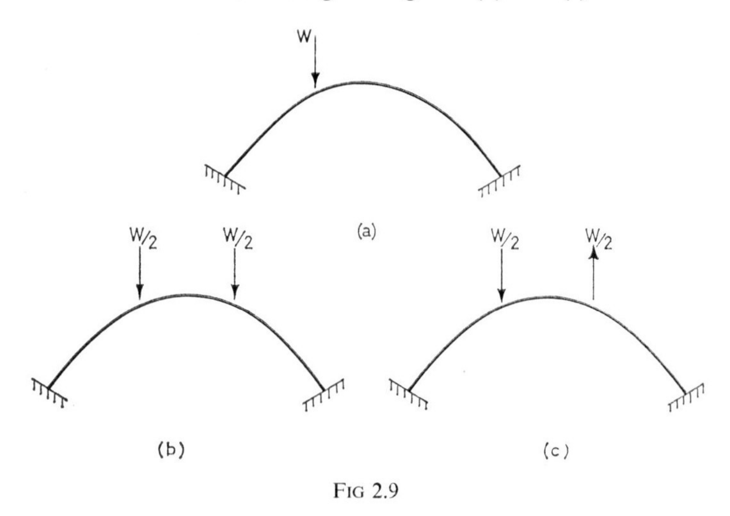

FIG 2.9

Analysis of the situation of Fig. 2.9(a) would involve determining three redundancies. However, the symmetrical situation depicted in Fig. 2.9(b) has only two redundancies, and in Fig. 2.9(c) the skew-symmetry reduces the problem to one redundancy. It would be simpler to solve these latter two problems, and then superpose the results, than to attempt a direct analysis of the original problem.

3
Virtual work and energy concepts

3.1. INTRODUCTION

The Principle of Virtual Work will be used in Chapters 4–6 as the basis for all of the indirect methods of structural analysis to be described. It will be shown in particular how all of the energy theorems can be derived directly from it and can therefore be regarded as special cases of the principle. It will also be used in Chapter 7 to prove the Reciprocal Theorems and in Chapter 8 to establish the various theorems of plastic and incremental collapse for framed structures. The principle thus occupies a central position in structural analysis, as realised for example by Lamb (1923). It is, however, often disregarded in the literature, and is sometimes quoted in certain rather special forms rather than in the general form which will be given here.

Following a derivation of the principle for the special case of pin-jointed plane trusses, the form taken by the principle for plane frames is quoted, the proof being given in an appendix. It is then shown that the principle can be used to transform an actual geometrical problem into a hypothetical equilibrium problem, or alternatively an actual equilibrium problem can be transformed into a hypothetical geometrical problem. Simple examples of these two types of transformation are given, although a detailed discussion of these transformations forms the subject matter of Chapters 4–6.

Finally, the chapter concludes with definitions of the three energy quantities which are used in the energy theorems, namely strain energy, complementary energy and potential energy.

3.2. PRINCIPLE OF VIRTUAL WORK

The appropriate axioms from which the Principle of Virtual Work may be derived, and even the precise nature of the principle itself, are matters for debate (Niles, 1943; Charlton, 1955). In what follows, only the elementary laws of statics and of geometry are assumed.

Proof for pin-jointed plane trusses

The principle is best explained and proved in relation to a particular form of structure, and the pin-jointed plane truss is convenient for this purpose.

The principle is concerned with sets of external loads and internal bar forces which satisfy all the requirements of equilibrium, and also with sets of joint deflections and bar extensions which satisfy all the requirements of compatibility. To these basic requirements must be added the further condition that gross deformations are excluded, so that the conditions of equilibrium and compatibility may be written down using the orientations of members in the undistorted structure.

A formal statement of the principle is as follows:

Let H_i and V_i denote the horizontal and vertical components of external load at a typical joint i, and let P_{ij} denote the tension in a typical bar ij. The external loads (H_i, V_i) and internal forces P_{ij} are presumed to satisfy all the requirements of equilibrium, and will be referred to collectively as the equilibrium system (H_i, V_i, P_{ij}).

Let h_i and v_i denote the horizontal and vertical components of deflection at a typical joint i, and let δl_{ij} denote the extension of a typical bar ij. The deflections (h_i, v_i) and bar extensions δl_{ij} are presumed to satisfy all the requirements of compatibility, and will be referred to collectively as the compatible system $(h_i, v_i, \delta l_{ij})$.

Then the Principle of Virtual Work states that

$$\sum_i (H_i h_i + V_i v_i) = \sum_{ij} P_{ij} \delta l_{ij}, \tag{3.1}$$

where the forces (H_i, V_i, P_{ij}) constitute an equilibrium system and the deformations $(h_i, v_i, \delta l_{ij})$ constitute a compatible system. The summations on the left- and right-hand sides of equation (3.1) cover all the joints i and bars ij, respectively.

This equation will usually be quoted in the following form:

$$\sum_{\text{joints}} (H^* h^{**} + V^* v^{**}) = \sum_{\text{bars}} P^* (\delta l)^{**} \tag{3.1a}$$

Here the asterisk notation emphasises that (H^*, V^*, P^*) is an equilibrium system and $(h^{**}, v^{**}, (\delta l)^{**})$ is a compatible system, and the suffix notation is dropped. However, the suffix notation is required when proving the result.

The Principle of Virtual Work is, of course, widely used in mechanics; its first statement in the above form for structural problems appears to have been given by Mohr (1874).

It will be appreciated from the nature of equation (3.1) that if the external loads (H_i, V_i) were actually applied to the truss, and the corresponding bar forces were P_{ij}, then if these loads and bar forces remained constant while the deformations took place, this equation would express the equality of the work done on the structure and the work absorbed in the bars. Even this situation is, of course, hypothetical in character, since it would require all the deformations to occur *after* the loads had been applied and the bar forces established. However, it is inadvisable to attach even this spurious degree of physical meaning to the principle. Equation (3.1) holds true even if the equilibrium system (H_i, V_i, P_{ij}) does not correspond to any actual loading on the structure, and if the compatible system $(h_i, v_i, \delta l_{ij})$ does not represent any actual deformations, and in particular those which would be caused by the loads. It is only necessary that the force and deformation systems should satisfy respectively the requirements of equilibrium and compatibility.

Despite these remarks, it is a useful aid to memory to note that equation (3.1) states that *The virtual work done by the loads*

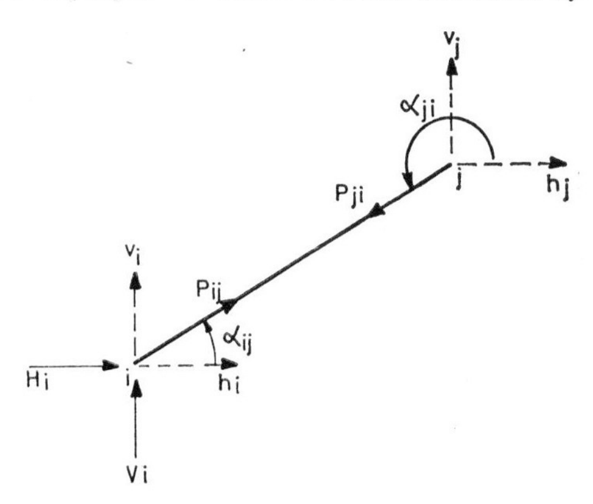

FIG 3.1

is equal to the virtual work absorbed in the bars, it being understood that this statement presumes constancy of the loads and bar forces, whether these be real or imaginary, while the deformations take place. This verbal statement of the principle is equally applicable to other types of structure.

The proof is readily established. Consider any typical bar *ij* in the truss which makes an angle α_{ij} with the horizontal, as shown in Fig. 3.1. It is, of course, presumed that α_{ij} is not affected appreciably by the deformations. The tension in this bar acting on the joint *i* is of course equal to the tension acting on the joint *j*, so that

$$P_{ij} = P_{ji} \tag{3.2}$$

Imagine that at the joint *i* there are external loads H_i and V_i in the horizontal and vertical directions, respectively. It is supposed that these loads and the internal bar forces P_{ij} satisfy the conditions of equilibrium, which for this joint are

$$\left.\begin{aligned} H_i + \sum_j P_{ij} \cos \alpha_{ij} &= 0 \\ V_i + \sum_j P_{ij} \sin \alpha_{ij} &= 0 \end{aligned}\right\}, \qquad (3.3)$$

the summations in each equation covering all the members radiating from joint i. Equations (3.3) express the requirements of statical equilibrium at the typical joint i.

Now imagine that the joints undergo small horizontal and vertical deflections, these being h_i and v_i, respectively, at joint i and h_j and v_j at joint j. The bars of the truss are at the same time supposed to undergo appropriate extensions such as δl_{ij} for bar ij so that they can continue to meet at the new positions of the joints and therefore satisfy the requirements of compatibility.

The extension δl_{ij} of the bar ij may be regarded as consisting of two components, one of which is attributable to the deflections of the joint i and is denoted by $(\delta l)'_{ij}$ and the other to the deflections of joint j and which is denoted by $(\delta l)''_{ji}$. These two components of δl_{ij} are given by

$$\left.\begin{aligned} (\delta l)'_{ij} &= -h_i \cos \alpha_{ij} - v_i \sin \alpha_{ij} \\ (\delta l)''_{ji} &= -h_j \cos \alpha_{ji} - v_j \sin \alpha_{ji} \end{aligned}\right\} \qquad (3.4)$$

and

$$\delta l_{ij} = (\delta l)'_{ij} + (\delta l)''_{ji} \qquad (3.5)$$

In the second of equations (3.4), α_{ji} is taken to be $(\pi + \alpha_{ij})$ for consistency of notation. Equations (3.4) and (3.5) express the requirements of geometrical compatibility.

Multiplying the first of equations (3.3) by h_i and the second by v_i, and adding, it follows that

$$H_i h_i + V_i v_i + \sum_j (h_i P_{ij} \cos \alpha_{ij} + v_i P_{ij} \sin \alpha_{ij}) = 0$$

Making use of the first of equations (3.4), it is then seen that

$$H_i h_i + V_i v_i = \sum_j P_{ij} (\delta l)'_{ij}$$

Both sides of this equation are now summed to cover all the joints in the truss. This presents no difficulty as far as the left-hand side is concerned. For the right-hand side it is noted that a summation covering all the joints must also cover each end of every bar comprising the truss. For instance, a term $P_{ij}(\delta l)'_{ij}$ has arisen in respect of bar ij due to a consideration of joint i. From joint j a term $P_{ji}(\delta l)''_{ji}$ will also arise for this bar. From equation (3.2) this term is equal to $P_{ij}(\delta l)''_{ji}$, and when the summation is carried out these two terms can be added to give $P_{ij}\delta l_{ij}$, from equation (3.5). Similar terms arise for every bar in the truss, so that finally equation (3.1) is obtained, namely

$$\sum_{i}(H_i h_i + V_i v_i) = \sum_{ij} P_{ij}\delta l_{ij}$$

In this equation, which expresses the Principle of Virtual Work, the summation on the left-hand side covers all the joints in the truss and the summation on the right-hand side covers all the bars.

It should be emphasised that for this result to hold true it is not necessary for the external loads to be actual loads on the structure, or for the bar forces to be those forces which would be induced by the loads. Nor is it necessary for the deflections and bar extensions to be those which would be caused by the loads. The two conditions which must be fulfilled are that the force system (H_i, V_i, P_{ij}) should satisfy all the requirements of equilibrium, and that the deformation system (h_i, v_i, δl_{ij}) should satisfy all the requirements of compatibility. In addition, it should be mentioned again that gross deformations are excluded.

An alternative, but less formal, proof is often given as follows. Provided that the conditions of equilibrium at each joint, as expressed by equations (3.3), are fulfilled, there is no resultant force on any joint. The net work done if the joint is imagined to undergo a small displacement is therefore zero. Summing this result for displacements of all the joints, it follows that the net work done on all the joints must be zero. This can be seen to be an alternative statement of equation (3.1).

Principle of Virtual Work for frames

For other types of structure the Principle of Virtual Work is readily written down by equating the virtual work done by the external loads to the virtual work absorbed in the members. For instance, in the case of plane frames, the principle takes the following form:

$$\sum_{\text{joints}} (H^*h^{**}+V^*v^{**}+C^*\phi^{**})+\oint (w_n^*y_n^{**}+w_t^*y_t^{**})\,ds$$
$$=\sum_{\text{hinges}} M^*\theta^{**}+\oint (M^*\kappa^{**}+P^*\varepsilon^{**})\,ds \qquad (3.6)$$

In this equation, the symbols H^*, V^*, h^{**} and v^{**} have the same significance as previously. C^* denotes an external couple applied to a joint, and ϕ^{**} is the joint rotation. w_n^* and w_t^* denote the intensities of normal and tangential loading on a member, and y_n^{**} and y_t^{**} are correspondingly the normal and tangential components of deflection. M^* denotes the bending moment at any section, and κ^{**} is correspondingly the curvature, while θ^{**} is the rotation at any hinge which may be present. As before, P^* is the axial tension at any section, and ε^{**} denotes the axial tensile strain.

The integrals are taken over the lengths s of all the members of the structure. The asterisk notation again emphasises that (H^*, V^*, C^*, w_n^*, w_t^*, M^*, P^*) is an equilibrium system and (h^{**}, v^{**}, ϕ^{**}, y_n^{**}, y_t^{**}, θ^{**}, κ^{**}, ε^{**}) is a compatible system.

The sign conventions for forces and displacements are consistent in the sense that the product of positive force and correspondingly positive displacement gives positive work.

Equation (3.6) is proved formally in Appendix A, the proof consisting of setting out the requirements of equilibrium and compatibility and then showing that provided these requirements are fulfilled, equation (3.6) is an identity.

3.3. VIRTUAL WORK TRANSFORMATIONS

As has been seen, the Principle of Virtual Work relates force systems which satisfy the requirements of equilibrium and deformation systems which satisfy the requirements of compatibility. In any particular application of the principle, the force system could either be the actual set of external loads and the corresponding internal forces, or some hypothetical force system which happens to satisfy the conditions of equilibrium. Similarly, the deformation system could consist either of the actual joint deflections and internal deformations of the structure, or some hypothetical deformation system which satisfies the conditions of compatibility. There are thus four possibilities, as follows:

	Force system	Deformation system
(i)	Actual	Actual
(ii)	Hypothetical	Actual
(iii)	Actual	Hypothetical
(iv)	Hypothetical	Hypothetical

Of these possibilities, (i) would reduce to an equation of real work, and is of no especial significance, and (iv) is rarely used. The chief applications are (ii) and (iii), and each of these two possibilities achieves a transformation. The nature of these transformations is best understood with reference to particular problems.

Transformation of an actual geometrical problem into a hypothetical equilibrium problem

Consider the statically determinate Warren girder illustrated in Fig. 3.2(a), and suppose that due to some loading, which need not be specified, all the forces in the bars have been determined. From known member characteristics for each bar it will also be supposed that the extension δl of each bar has also been found; this extension might be due solely to the loading or possibly due to some additional influence such as a change of temperature. There then remains the purely geometrical problem of determining the deflections of the various joints. This could be solved by the construction of a Williot-Mohr diagram, but this procedure is tedious even for the simple example shown. There is therefore a strong incentive to avoid the geometrical issue, and it will now be shown that by using the Principle of Virtual Work it is possible to transform this actual geometrical problem into a hypothetical equilibrium problem which is much simpler to solve.

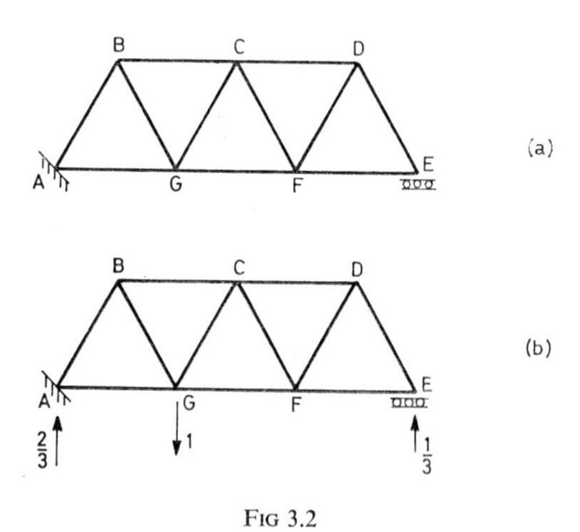

FIG 3.2

The device which is employed is to use the actual bar extensions and joint deflections in the virtual work equation, since these are known to satisfy the requirements of compatibility. In addition, a hypothetical equilibrium system of forces is used which produces the required result. For instance, suppose that the vertical deflection v_G of G is sought. The equilibrium system which is used is the set of bar forces which satisfies the requirements of equilibrium with a hypothetical (or dummy) unit vertical load at G. These forces are readily found by resolving at each joint in turn, once the reactions at A and E are determined, as in Fig. 3.2(b), and are shown in Table 3.1, in which the extensions of the bars are also given.

The virtual work equation for this type of structure is equation (3.1a), namely

$$\sum_{\text{joints}} (H^*h^{**} + V^*v^{**}) = \sum_{\text{bars}} P^*(\delta l)^{**}$$

Considering the left-hand side of this equation first, it is seen that at joint A, h_A and v_A are constrained to be zero, so that there is no contribution to the summation from this joint. At joint E, v_E is zero and H_E is zero, so that again there is no contribution. At joints B, C, D and F, the external loads are zero, and so the only non-zero term in the summation arises from the joint G. The above equation thus becomes

$$1.v_G = \sum_{\text{bars}} P'\delta l$$

where P' is the force in a typical bar due to the hypothetical external loads of Fig. 3.2(b).

The data are summarised in Table 3.1.

TABLE 3.1

Member	AG	GF	FE	BC	CD	AB
δl (in.)	0·10	0·10	0·10	−0·05	−0·05	−0·05
p'	$\frac{2}{3\sqrt{3}}$	$\frac{1}{\sqrt{3}}$	$\frac{1}{3\sqrt{3}}$	$-\frac{4}{3\sqrt{3}}$	$-\frac{2}{3\sqrt{3}}$	$-\frac{4}{3\sqrt{3}}$

Member	BG	GC	CF	FD	DE
δl (in.)	0·10	0	0	0·10	−0·05
p'	$\frac{4}{3\sqrt{3}}$	$\frac{2}{3\sqrt{3}}$	$-\frac{2}{3\sqrt{3}}$	$\frac{2}{3\sqrt{3}}$	$-\frac{2}{3\sqrt{3}}$

From Table 3.1, it is readily found that $\Sigma P'\delta l = 0{\cdot}346$ in., and this is therefore the value of v_G.

It will be appreciated that by this application of the Principle of Virtual Work, the actual geometrical problem of determining v_G has been transformed into the equilibrium problem of determining the bar forces which satisfy the requirements of equilibrium with the hypothetical loading of Fig. 3.2(b). The procedure used, although related to a particular problem, is quite general; it was first developed by Maxwell (1864) and independently by Mohr (1874). A more detailed discussion of this type of transformation, which is termed the Maxwell–Mohr or "dummy unit load" method, will be given in Chapter 5.

Transformation of an actual equilibrium problem into a hypothetical geometrical problem

The other type of virtual work transformation will be explained with reference to the portal frame whose dimensions and loading are as illustrated in Fig. 3.3(a). It will be supposed

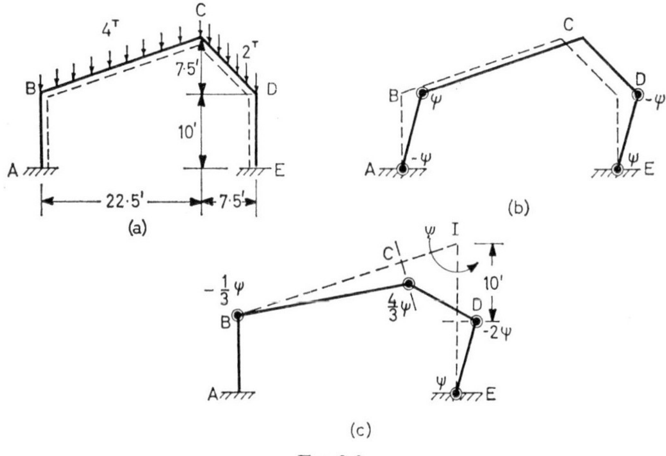

FIG 3.3

that it is required to find the equations of equilibrium which must be obeyed by the bending moments at A, B, C, D and E.

The first step is to decide how many independent equations of equilibrium there must be. This is done by noting that the number of unknown bending moments is 5, and the number of redundancies is 3. The number of independent equations of equilibrium is therefore two, since if there are two equations connecting the five bending moments, a knowledge of any three of these bending moments would enable the remaining two moments to be determined from these equations.

The two equations of equilibrium may be derived from the Principle of Virtual Work by using the actual external loads and internal forces and moments, since these must satisfy the requirements of equilibrium. In addition, two hypothetical deformation systems are used in turn, each of which satisfies all the requirements of compatibility. These are the two mechanism motions illustrated in Figs. 3.3(b) and (c), each of which is characterised by the presence of four hinges, which permit the movements illustrated while the members between the joints remain straight.

In each mechanism motion the curvature at every section is zero, and no extensions or contractions of the members are involved, so that the internal deformations consist merely of the rotations at the four hinges. These rotations must be given signs consistent with the bending moment convention. Positive bending moments will be defined as causing tension in the fibres of the members adjacent to the dotted line in Fig. 3.3(a), and so the signs of the hinge rotations in the two mechanisms will be as shown in Figs. 3.3(b) and (c).

The Principle of Virtual Work for plane frames was given in equation (3.6) as

$$\sum_{\text{joints}} (H^*h^{**}+V^*v^{**}+C^*\phi^{**})+\oint (w_n^* y_n^{**}+w_t^* y_t^{**})\,\mathrm{d}s$$

$$=\sum_{\text{hinges}} M^*\theta^{**}+\oint (M^*\kappa^{**}+P^*\varepsilon^{**})\,\mathrm{d}s$$

In this particular problem, the external load system involves no concentrated loads or couples, so that H^*, V^* and C^* are zero at every joint. The hypothetical deformation systems involve no curvatures or extensions of the members, so that κ^{**} and ε^{**} are also zero everywhere. The virtual work equation thus reduces to

$$\oint (w_n^* y_n^{**}+w_t^* y_t^{**})\,\mathrm{d}s=\sum_{\text{hinges}} M^*\theta^{**} \tag{3.7}$$

Considering first the sidesway mechanism of Fig. 3.3(b), it will be seen that the magnitudes of the hinge rotations are all the same, since in this mechanism the roof BCD remains rigid. The horizontal movement at B and at D is 10ψ ft, and any point on the roof BCD will also move horizontally through this same distance. Since the only loading on the roof is vertical, $(w_n^* y_n^{**} + w_t^* y_t^{**})$ is everywhere zero, so that the virtual work done by the loads during this mechanism movement is zero. It follows from equation (3.7) that

$$\sum_{\text{hinges}} M^*\theta^{**} = 0$$

$$M_A(-\psi)+M_B(\psi)+M_D(-\psi)+M_E(\psi) = 0$$

$$-M_A+M_B-M_D+M_E = 0 \qquad (3.8)$$

This is one of the two equations of equilibrium.

The mechanism motion of Fig. 3.3(c) is somewhat more difficult to analyse. The member AB remains vertical, so that BC rotates about B. The apex C must therefore move in a direction perpendicular to BC. Since ED rotates about E, D must move horizontally. The directions in which the ends C and D of the member CD move are therefore known, and this member must rotate about an instantaneous centre I at the position shown in the figure.

Presuming a rotation ψ of CD about I, it follows that the movements of C and D are respectively $(IC)\psi$ and $(ID)\psi$. The hinge rotations at B and E are thus $\psi IC/BC = \frac{1}{3}\psi$, and $\psi ID/ED = \psi$, respectively. The hinge rotation at C is then seen to be of magnitude $\psi+\frac{1}{3}\psi = \frac{4}{3}\psi$, and at D the magnitude of the rotation is $\psi+\psi = 2\psi$.

To evaluate the virtual work done by the external loads, it is noted that the vertical movement of C is $7{\cdot}5\psi$ ft. Since B does not move vertically, and the member BC remains straight, the uniformly distributed 4 ton load does virtual work $0{\cdot}5\times 4 \times 7{\cdot}5\psi = 15\psi$ ton ft. Similarly, since D does not move vertically, the 2 ton vertical load on CD does virtual work $0{\cdot}5\times 2 \times 7{\cdot}5\psi = 7{\cdot}5\psi$ ton ft.

The total virtual work done by the loads is therefore $22{\cdot}5\psi$ ton ft. Equating this to the work absorbed at the hinges, it follows that

$$M_B(-\tfrac{1}{3}\psi)+M_C(\tfrac{4}{3}\psi)+M_D(-2\psi)+M_E(\psi) = 22{\cdot}5\psi$$

$$-\tfrac{1}{3}M_B+\tfrac{4}{3}M_C-2M_D+M_E = 22{\cdot}5 \quad (3.9)$$

It will be appreciated that equations (3.8) and (3.9) could have been derived directly from statical considerations. However, the above derivation is perhaps simpler. Further, in a

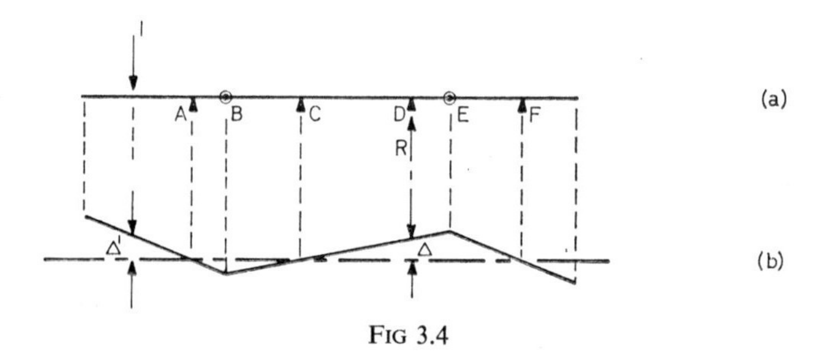

FIG 3.4

plastic collapse analysis it would be necessary to analyse the mechanisms of Figs. 3.3(b) and (c) as possible collapse mechanisms, and so the results of the geometrical analysis would already be available. This is also true if an elastic moment distribution analysis was to be carried out: in this case the mechanisms define the possible relative movements of the joints in the sway deformations of the frame (Neal, 1957).

A further illustration of this type of transformation is afforded by the statically determinate beam system of Fig. 3.4(a). It is supposed that it is required to construct the influence line for the variation in the reaction R at D as a unit load crosses the span. This could be dealt with by statics, but it is more convenient to use the hypothetical deformation system depicted in Fig. 3.4(b). This consists of a vertical displacement Δ of the support D, with the other supports remaining in their original positions and each segment of the beam remaining straight. The deflection at the position of the unit load is shown as $-\Delta'$.

The equation of virtual work is simply

$$R\Delta - 1.\Delta' = 0$$

It follows that if the deformation system is drawn to a scale which makes Δ unity, $R = \Delta'$, so that the ordinates of displacement are a direct measure of R and therefore constitute the required influence line.

3.4. STRAIN ENERGY

The remainder of this chapter will be concerned with the definition of three energy quantities which are widely used in the various theorems of structural analysis. These quantities are *strain energy*, *complementary energy* and *potential energy*. The present section will be concerned with strain energy.

Trusses

Consider first a bar which may form part of a pin-jointed truss. In general, the length l of the bar will depend upon various factors such as the axial force in the bar, its temperature, and other environmental quantities. A length l_0 may be selected for the datum conditions of zero axial force and certain values of the other environmental factors such as average room temperature. Then in any other conditions the length l will be defined by

$$l = l_0 + \delta l \qquad (3.10)$$

If all the environmental conditions except axial force are kept at their datum values, the change in length of the bar will be due solely to the axial tension P. Under these circumstances, the symbol e will be used to denote the extension of the bar, so that

$$l = l_0 + e \qquad (3.11$$

It will be supposed that there is a unique relationship between the tension P in the bar and its extension e, as shown in Fig. 3.5. Then if the bar has extended to the point A, the strain energy u is defined as

$$u = \int_0^e P\,\mathrm{d}e \tag{3.12}$$

Thus the strain energy corresponds to the cross-hatched area in Fig. 3.5.

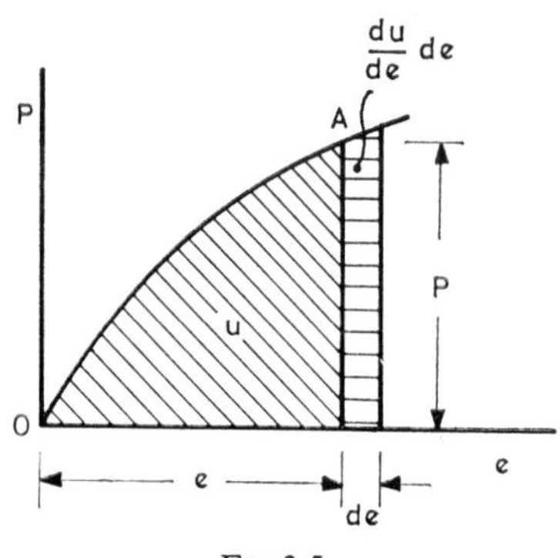

FIG 3.5

A physical meaning is often attached to this quantity; it represents the work done on the bar in extending it from O to A, and if the (P,e) relation is reversible this would also be the strain energy stored. However, the usefulness of strain energy is not confined to cases in which the (P,e) relation is reversible. The definition of equation (3.12) is still valid if the loading and unloading characteristics of the member differ, provided that unloading of the member does not actually occur in the problem being analysed. Thus, it is advisable not to regard the strain energy as a physically significant quantity, but instead to treat it as a quantity which happens to be useful in structural analysis and which is defined by equation (3.12).

The value of the derivative $\mathrm{d}u/\mathrm{d}e$ will be required later in Chapter 6. It will be seen from Fig. 3.5 that if the bar extension is increased from e to $e+\mathrm{d}e$, u increases by an amount $P\,\mathrm{d}e$. It follows that

$$\frac{\mathrm{d}u}{\mathrm{d}e}\,\mathrm{d}e = P\mathrm{d}e$$

so that

$$\frac{\mathrm{d}u}{\mathrm{d}e} = P \tag{3.13}$$

For a complete structure the strain energy U is the sum of the strain energies for each of the bars, so that

$$U = \sum_{\text{bars}} \int_0^e P\,\mathrm{d}e \tag{3.14}$$

This presumes that the supports are either rigid, as in Fig. 3.6(a), or free, as in Fig. 3.6(b). In Fig. 3.6(a) the joint displacement is zero, so that the reactions H and V do no work; in

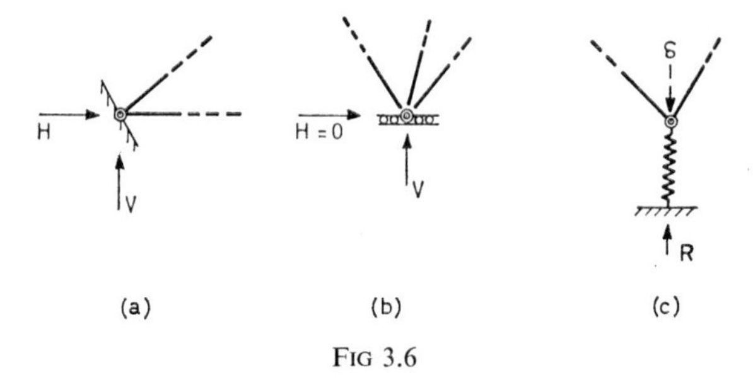

FIG 3.6

Fig. 3.6(b) the vertical displacement is zero, so that V does no work, and the roller support permits horizontal displacement but the horizontal reaction H is zero.

For the case of a support which permits some displacement under load, as depicted in Fig. 3.6(c), the strain energy U should include a term $\int R\,\mathrm{d}\delta$, where R is the reaction and δ is the corresponding joint displacement. However, this can be regarded as included in the summation of equation (3.14) by treating the spring as an extra bar in the truss.

Beams and frames

For beams and frames it is supposed that the curvature change at any section consists of a change κ due to the bending moment M together with a change κ_0 due to all other environmental changes. If δu denotes the strain energy for a length δs of the member,

$$\frac{\mathrm{d}u}{\mathrm{d}s} = \int_0^\kappa M \,\mathrm{d}\kappa, \tag{3.15}$$

so that $\mathrm{d}u/\mathrm{d}s$ is given by the cross-hatched area in Fig. 3.7.

The strain energy u for a member connecting two joints i and j is then given by

$$u = \int_i^j \frac{\mathrm{d}u}{\mathrm{d}s} \,\mathrm{d}s, \tag{3.16}$$

and for a complete structure the strain energy U is obtained by summing u over all the members.

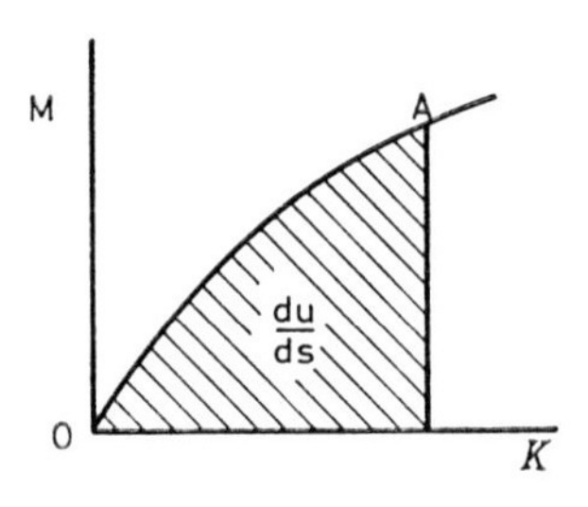

FIG 3.7

If there are supports which permit displacement under load, as in Fig. 3.6(c), the strain energy U should include terms such as $\int R \,\mathrm{d}\delta$, as was the case for trusses. Further, if the joint between a member and a rigid support, or between two members, permits a relative rotation θ due to a joint moment M, a term $\int M \,\mathrm{d}\theta$ must be included. The sign conventions for such terms present little difficulty in practice if it is remembered that U must represent the work done on all the components of the structure in producing their deformations.

3.5. COMPLEMENTARY ENERGY

Trusses

As opposed to strain energy, complementary energy involves the total extension δl of each bar due to all causes.

It is convenient to regard this total extension as made up of two parts, the first being the extension e due to the axial

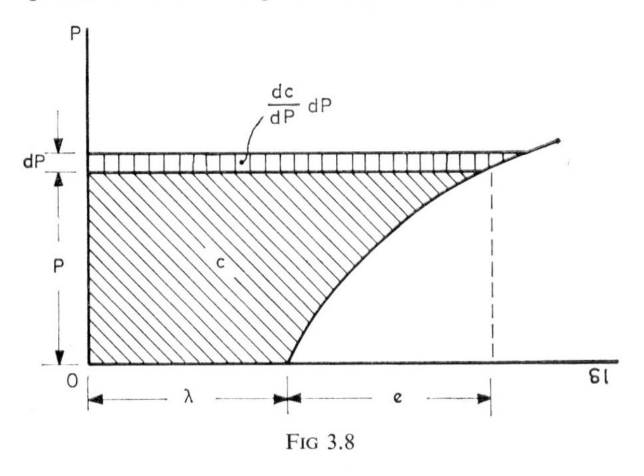

FIG 3.8

tension P, and the second being the remaining extension λ due to all other environmental changes such as temperature, etc. Thus if the bar, of datum length l_0, changes its length to $l_0+\delta l$,

$$\delta l = e+\lambda \tag{3.17}$$

The complementary energy c for a typical bar may now be defined with reference to Fig. 3.8 as

$$c = P\lambda+\int_0^P e\,dP, \tag{3.18}$$

so that c corresponds to the cross-hatched area in this figure.

Complementary energy has no physical meaning, and is merely defined because it is a useful quantity in structural analysis. It should also be emphasised that Fig. 3.8 is purely a formal device for defining c, and does not necessarily represent the behaviour of an actual bar in a real physical problem. For instance, consider a redundant truss in which one of the redundant members is cooled while the others remain at datum temperature, while no other environmental changes occur and

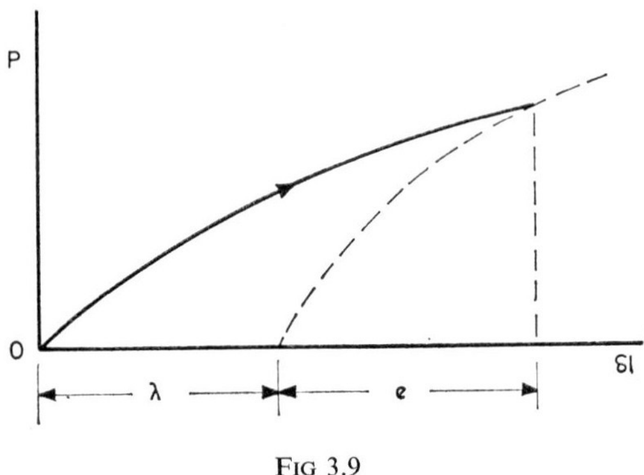

FIG 3.9

no loads are applied. Under these circumstances, the tension in the redundant bar would increase steadily with drop in temperature, so that λ and P would increase simultaneously. The $(P, \delta l)$ relation for the member would therefore be as depicted in Fig. 3.9.

Figure 3.8 is in fact best thought of as defining an energy quantity complementary to the strain energy u in the sense that

$$u+c = P(\lambda+e) = P\delta l \tag{3.19}$$

The usefulness of complementary energy, like strain energy, is restricted to cases in which the relationship between P and e is unique. The definitions of equations (3.18) and (3.19) are, however, applicable to non-linear (P,e) relations, and even, to cases in which the loading and unloading characteristics are different, provided of course that unloading does not occur in the problem under consideration.

The derivative $\mathrm{d}c/\mathrm{d}P$ will be required later in Chapters 4 and 5. It will be seen from Fig. 3.8 that

$$\frac{\mathrm{d}c}{\mathrm{d}P}\,\mathrm{d}P = (\lambda+e)\,\mathrm{d}P$$

so that

$$\frac{\mathrm{d}c}{\mathrm{d}P} = \lambda+e \tag{3.20}$$

For a complete structure, the complementary energy C is given by

$$C = \sum_{\text{bars}}\left[P\lambda+\int_0^P e\,\mathrm{d}P\right], \tag{3.21}$$

and flexible supports are treated as additional bars as in the definition of U.

Beams and frames

The corresponding definition of complementary energy for a beam or frame member is

$$\frac{\mathrm{d}c}{\mathrm{d}s} = M\kappa_0+\int_0^M \kappa\,\mathrm{d}M, \tag{3.22}$$

where κ and κ_0 are as defined previously.

The complementary energy c for a member connecting two joints i and j is

$$c = \int_i^j \frac{\mathrm{d}c}{\mathrm{d}s}\,\mathrm{d}s, \tag{3.23}$$

and for a complete structure C is obtained by summing over all the members.

Terms for flexible supports and joints should also be included when these occur.

3.6. TOTAL POTENTIAL

The third energy quantity which will be required later is the total potential ϕ. Its definition is simply

$$\phi = U+V, \tag{3.24}$$

where U is the strain energy for the structure and V represents the potential energy of the external loads. For trusses, V is given by the summation

$$V = -\sum_{\text{joints}} (Hh+Vv), \tag{3.25}$$

where h and v are the horizontal and vertical components of deflection at a joint, measured from a specified datum configuration for the truss.

The corresponding definition for frames is

$$V = -\sum_{\text{joints}} (Hh+Vv+C\phi)-\oint (w_n y_n+w_t y_t)\,\mathrm{d}s \tag{3.26}$$

4

Indeterminate structures by the compatibility method

4.1. INTRODUCTION

In Chapter 1 it was pointed out that the analysis of a structure can be carried out either by the compatibility method, using force variables, or the equilibrium method, using deformation variables. The present chapter is concerned with the former method, in which the primary objective is the determination of the redundancies of the problem. These are found by setting up and solving the equations of compatibilty. By contrast, the equilibrium method has as its primary objective the determination of the unknown components of deformation, which are found from the equations of equilibrium.

The compatibility method is generally used when the number of redundancies is smaller than the number of unknown deformation components, although the advent of computer programmes for structural analysis has made the choice of method less obvious (Livesley, 1964). Most practical trusses fall into this category, and furthermore the compatibility method is most readily explained in relation to these structures. The earlier examples in this chapter therefore deal exclusively with trusses, and other types of structure are discussed at the end of the chapter.

Each problem to be discussed is analysed by an indirect method, so that the equations of compatibility are not set up by means of geometrical arguments. Instead, the Principle of Virtual Work is used to transform the actual geometrical problem of establishing the conditions of compatibility into a hypothetical equilibrium problem.

Certain energy theorems of structural analysis are also shown to achieve identical transformations, and are discussed in relation to those problems to which they are applicable. These theorems are

Engesser's Theorem of Compatibility, and
Castigliano's Theorem of Compatibility.

The nomenclature used to describe these and other energy theorems is that recommended by Matheson (1959).

4.2. TRUSS WITH SINGLE REDUNDANCY

The first problem which will be considered is the truss illustrated in Fig. 4.1(a). All the bars are of length 5 ft or $5\sqrt{2}$ ft. The relation between stress σ(ton/in^2) and strain ε for each bar is

$$\varepsilon = \frac{\sigma}{4500} + 10^{-6}\sigma^3 \tag{4.1}$$

This stress/strain relation is illustrated in Fig. 4.2; it will be seen that it corresponds to a 0·1 per cent. proof stress of 10 ton/in^2. The cross-sectional area A_0 of each bar is given below in Table 4.1. It is required to find the forces in all the bars due to the application of the vertical load of 10 ton at C, the truss being free from stress when unloaded.

This truss has one redundancy, whereas there are initially six unknown components of deformation, namely two components at each of the three joints B, C and D. It is therefore more convenient to work with force variables.

The redundant bar can be chosen arbitrarily as any of the bars AB, BD, DE, AD or BE; the forces in BC and CD can be determined by statics alone. For the subsequent analysis, the tension R in bar AD will be taken as the single redundancy; the statically determinate truss which remains when AD is omitted is of course the basic truss. The analysis is then directed towards the determination of R in the first instance.

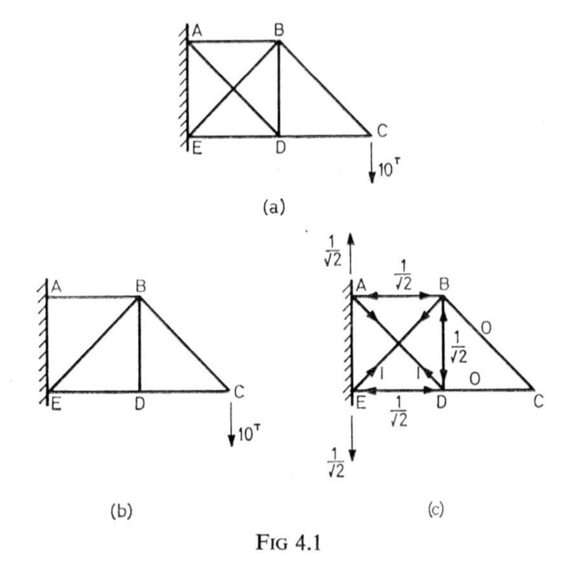

FIG 4.1

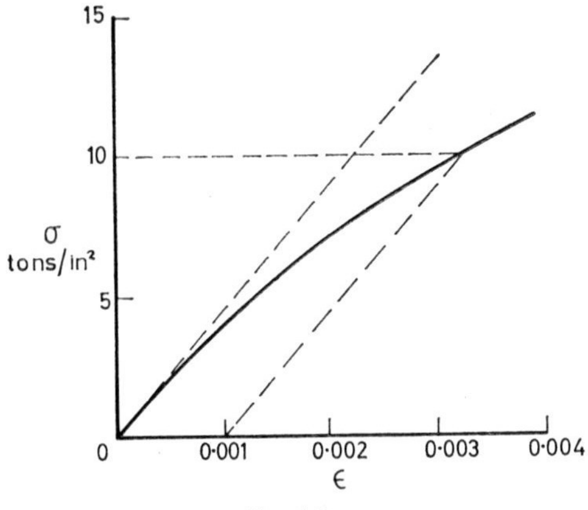

FIG 4.2

The compatibility equation for the determination of R will first be derived using the Principle of Virtual Work, and it will then be shown that this equation can also be obtained by applying Engesser's Theorem of Compatibility.

Solution by virtual work

Equilibrium

The first step in the analysis is to express the conditions of equilibrium. This is done most conveniently by establishing two force systems. The first of these is the set of bar forces in equilibrium with the vertical load of 10 ton at C when the force in AD is zero, as in Fig. 4.1(b). These forces will be denoted by P_W. The second is the set of forces in equilibrium with a unit tension in AD with no load at C, as in Fig. 4.1(c). These forces will be denoted by r. Then the actual force P in any bar is found by the Principle of Superposition for forces to be

$$P = P_W + rR \tag{4.2}$$

The force systems P_W and r are given in Table 4.1, together with the values of P derived from equation (4.2).

TABLE 4.1

Member	$P_W(ton)$	r	$P(ton)$	$A_0(in^2)$	$\sigma(ton/in^2)$
AB	20	$-1/\sqrt{2}$	$20-R/\sqrt{2}$	1·5	$13{\cdot}33-0{\cdot}471R$
BC	$10\sqrt{2}$	0	$10\sqrt{2}$	1·5	9·43
CD	-10	0	-10	1	-10
DE	-10	$-1/\sqrt{2}$	$-10-R/\sqrt{2}$	1·5	$-6{\cdot}67-0{\cdot}471R$
BD	0	$-1/\sqrt{2}$	$-R/\sqrt{2}$	1	$-0{\cdot}707R$
BE	$-10\sqrt{2}$	1	$-10\sqrt{2}+R$	1	$-14{\cdot}14+R$
AD	0	1	R	1	R

Stress/strain relation

Having expressed the force P in each bar in terms of R alone, the next step is to use the stress/strain relation to express the extension e of each bar in terms of R. From equation (4.1), it follows that

$$e = l_0\varepsilon = l_0\left[\frac{\sigma}{4500}+10^{-6}\sigma^3\right] \tag{4.3}$$

where l_0(in.) is the original length of the bar and e(in.) is its extension due to the application of the 10 ton load.

Compatibility

Finally, there remains the geometrical problem of determining the correct value of R, which is such that the extension of AD due to the force R is equal to the separation of the points A and D in the basic truss. The Principle of Virtual Work will be used to transform this geometrical problem into a hypothetical equilibrium problem.

The transformation is achieved by using the actual bar extensions e in the Principle of Virtual Work, since these are known to satisfy the requirements of compatibility. The force system which is used is the system of Fig. 4.1(c), namely the forces r in the bars which satisfy all the requirements of equilibrium with a unit tension in AD.

The virtual work equation was given in Chapter 3, equation (3.1a), in the form

$$\sum_{\text{joints}} (H^*h^{**}+V^*v^{**}) = \sum_{\text{bars}} P^*(\delta l)^{**}$$

It will be seen that in the force system of Fig. 4.1(c) there are no external forces except at A and E, where the displacements are zero. Thus the terms on the left-hand side of the virtual work equation, representing the virtual work done by the (hypothetical) external loads, are all zero. The equation therefore becomes

$$\sum_{\text{bars}} P^*(\delta l)^{**} = \sum_{\text{bars}} re = 0 \tag{4.4}$$

It should be noted that the summation in this equation covers all the bars of the truss, and not just the basic truss with AD omitted.

Making use of equation (4.3), this becomes

$$\sum_{\text{bars}} r l_0 \left[\frac{\sigma}{4500} + 10^{-6}\sigma^3 \right] = 0 \qquad (4.5)$$

The data required for equation (4.4) or (4.5) are set out in Table 4.2.

TABLE 4.2

Member	l_0(*in.*)	*re*
AB	60	$-\frac{1}{\sqrt{2}} 60\left[\frac{(13 \cdot 33 - 0 \cdot 471R)}{4500} + 10^{-6}(13 \cdot 33 - 0 \cdot 471R)^3\right]$
BC	$60\sqrt{2}$	0
CD	60	0
DE	60	$-\frac{1}{\sqrt{2}} 60\left[\frac{(-6 \cdot 67 - 0 \cdot 471R)}{4500} + 10^{-6}(-6 \cdot 67 - 0 \cdot 471R)^3\right]$
BD	60	$-\frac{1}{\sqrt{2}} 60\left[\frac{-0 \cdot 707R}{4500} + 10^{-6}(-0 \cdot 707R)^3\right]$
BE	$60\sqrt{2}$	$60\sqrt{2}\left[\frac{(-14 \cdot 14 + R)}{4500} + 10^{-6}(-14 \cdot 14 + R)^3\right]$
AD	$60\sqrt{2}$	$60\sqrt{2}\left[\frac{R}{4500} + 10^{-6}R^3\right]$

From Table 4.2 it is found that equation (4.4) is

$$-\frac{60}{4500\sqrt{2}}[34 \cdot 94 - 5 \cdot 649R]$$

$$-\frac{60}{\sqrt{2}} 10^{-6} [(13 \cdot 33 - 0 \cdot 471R)^3 - (6 \cdot 67 + 0 \cdot 471R)^3$$
$$- (0 \cdot 707R)^3 + 2(14 \cdot 14 - R)^3 - 2R^3] = 0$$

This is the compatibility equation for the determination of R. The solution is found to be $R = 6 \cdot 49$ ton.

To complete the analysis the forces in the bars are tabulated, making use of Table 4.1. As a check, the stresses σ are also determined, and the extensions then evaluated from equation

(4.3). A direct check on equation (4.4) is then made in Table 4.3, from which it is seen that $\Sigma\, re = 0$ to within the limits of accuracy of the calculation.

TABLE 4.3

Member	*P(ton)*	*σ(ton/in²)*	*e(in.)*	*r*	*re*
AB	15·41	10·27	0·202	$-1/\sqrt{2}$	−0·143
BC	14·14	9·43	0·249	0	0
CD	−10	−10	−0·193	0	0
DE	−14·59	−9·72	−0·185	$-1/\sqrt{2}$	0·131
BD	−4·59	−4·59	−0·067	$-1/\sqrt{2}$	0·047
BE	−7·65	−7·65	−0·182	1	−0·182
AD	6·49	6·49	0·145	1	0·145
				$\Sigma\, re$	−0·002

Solution by Engesser's Theorem of Compatibility

Precisely the same calculations are performed if Engesser's Theorem of Compatibility is used (Engesser, 1889). This theorem states that if there are N redundancies R_1, R_2, . . . , R_q, . . . , R_n, and the complementary energy C is expressed in terms of these redundancies, then the N compatibility equations are given by

$$\frac{\partial C}{\partial R_1} = 0, \quad \frac{\partial C}{\partial R_2} = 0, \quad \ldots, \quad \frac{\partial C}{\partial R_q} = 0, \quad \ldots, \quad \frac{\partial C}{\partial R_n} = 0$$

For this particular truss, there is only one redundancy R, so that Engesser's Theorem of Compatibility states that the compatibility equation is

$$\frac{\partial C}{\partial R} = 0$$

The complementary energy c for any bar was defined by equation (3.18) as

$$c = P\lambda + \int_0^P e\, \mathrm{d}P$$

In this example λ, the change in length of the bar which is not due to P, is zero, so that

$$c = \int_0^P e\,\mathrm{d}P$$

Putting $P = A_0\sigma$, and substituting for e in terms of σ by equation (4.3),

$$\begin{aligned} c &= A_0 l_0 \int_0^\sigma \left[\frac{\sigma}{4500} + 10^{-6}\sigma^3\right] \mathrm{d}\sigma \\ &= A_0 l_0 \left[\frac{\sigma^2}{9000} + \tfrac{1}{4}10^{-6}\sigma^4\right] \end{aligned}$$

For each bar P, and therefore σ, can be expressed as a linear function of R by means of equation (4.2), so that

$$\sigma = \frac{P}{A_0} = \frac{P_W + rR}{A_0}$$

and

$$\frac{\mathrm{d}\sigma}{\mathrm{d}R} = \frac{r}{A_0}$$

It therefore follows that

$$\frac{\partial c}{\partial R} = \frac{\mathrm{d}c}{\mathrm{d}\sigma}\cdot\frac{\mathrm{d}\sigma}{\mathrm{d}R} = A_0 l_0 \left[\frac{\sigma}{4500} + 10^{-6}\sigma^3\right]\frac{r}{A_0}$$

Comparing this with equation (4.3), it is seen that

$$\frac{\partial c}{\partial R} = re$$

The total complementary energy C for the whole truss is found by summing c over the bars. It follows at once that

$$\frac{\partial C}{\partial R} = \sum_{\text{bars}} \frac{\partial c}{\partial R} = \sum_{\text{bars}} re$$

Engesser's Theorem of Compatibility therefore states that in this case the compatibility equation is given by

$$\frac{\partial C}{\partial R} = \sum_{\text{bars}} re = 0,$$

and this is precisely equation (4.4) which was derived by virtual work. Thus Engesser's Theorem of Compatibility in this instance achieves precisely the same transformation as the virtual work approach which was used to solve this problem.

4.3. TRUSS WITH SEVERAL REDUNDANCIES

A more complex problem is illustrated in Fig. 4.3. All the bars are originally of length L or $L\sqrt{2}$, and the relation

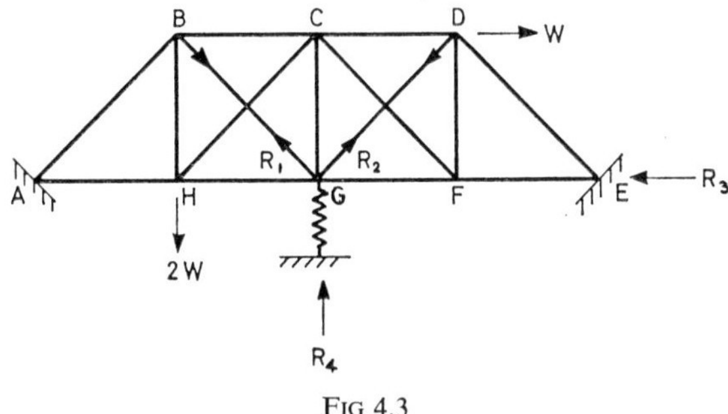

FIG 4.3

between tension P and extension e due to tension for each bar is

$$P = \mu_0 e$$

At A and E the truss is pinned to rigid supports, and at G it rests on an elastic footing with spring constant $0{\cdot}5\mu_0$, so that the downwards vertical displacement v_G of G is related to the reaction R_4 by

$$R_4 = 0{\cdot}5\mu_0 v_G$$

It is required to find the forces in the bars caused by the loading shown together with a temperature rise T above a datum temperature at which the truss is free from stress when unloaded. The coefficient of linear expansion for each bar is α.

The truss has four redundancies, which may be selected as the tensions R_1 and R_2 in bars BG and DG, respectively, the horizontal reaction R_3 at E, and the vertical reaction R_4 at G. The use of these four force variables is clearly advantageous for this problem since the number of deformation variables would be twelve.

Solution by virtual work

Equilibrium

As before, the analysis begins by considering the conditions of equilibrium. This involves investigating five force systems. In the first place, the forces P_W in the bars which would occur in the basic truss with the four redundancies R_1, R_2, R_3 and

TABLE 4.4

Member	P_W	r_1	r_2	r_3	r_4	l_0	μ
AB	$-1{\cdot}25\sqrt{2}.W$	0	0	0	$0{\cdot}5\sqrt{2}$	$\sqrt{2}.L$	μ_0
BC	$-1{\cdot}25W$	$-0{\cdot}5\sqrt{2}$	0	0	$0{\cdot}5$	L	μ_0
CD	$0{\cdot}25W$	0	$-0{\cdot}5\sqrt{2}$	0	$0{\cdot}5$	L	μ_0
DE	$-0{\cdot}75\sqrt{2}.W$	0	0	0	$0{\cdot}5\sqrt{2}$	$\sqrt{2}.L$	μ_0
AH	$2{\cdot}25W$	0	0	-1	$-0{\cdot}5$	L	μ_0
HG	$1{\cdot}5W$	$-0{\cdot}5\sqrt{2}$	0	-1	-1	L	μ_0
GF	$1{\cdot}5W$	0	$-0{\cdot}5\sqrt{2}$	-1	-1	L	μ_0
FE	$0{\cdot}75W$	0	0	-1	$-0{\cdot}5$	L	μ_0
BH	$1{\cdot}25W$	$-0{\cdot}5\sqrt{2}$	0	0	$-0{\cdot}5$	L	μ_0
CG	0	$-0{\cdot}5\sqrt{2}$	$-0{\cdot}5\sqrt{2}$	0	-1	L	μ_0
DF	$0{\cdot}75W$	0	$-0{\cdot}5\sqrt{2}$	0	$-0{\cdot}5$	L	μ_0
BG	0	1	0	0	0	$\sqrt{2}.L$	μ_0
CH	$0{\cdot}75\sqrt{2}.W$	1	0	0	$0{\cdot}5\sqrt{2}$	$\sqrt{2}.L$	μ_0
DG	0	0	1	0	0	$\sqrt{2}.L$	μ_0
CF	$-0{\cdot}75\sqrt{2}.W$	0	1	0	$0{\cdot}5\sqrt{2}$	$\sqrt{2}.L$	μ_0
G	0	0	0	0	-1	0	$0{\cdot}5\mu_0$

R_4 all zero, and with the external loads applied, are found. Then with zero external loads and $R_2 = R_3 = R_4 = 0$, the forces r_1 due to $R_1 = 1$ are determined. Similarly, the forces r_2 due to $R_2 = 1$, with zero external loads and $R_1 = R_3 = R_4 = 0$ are found, and two other systems r_3 and r_4 are similarly defined.

The actual force P in any bar is then given by the Principle of Superposition of forces as

$$P = P_W + r_1 R_1 + r_2 R_2 + r_3 R_3 + r_4 R_4 \tag{4.6}$$

These force systems are given in Table 4.4.

Member characteristics

The next step in the analysis is to use the member characteristics to determine the extension δl of each member. This is given by

$$\delta l = \lambda + e = \alpha T l_0 + P/\mu \tag{4.7}$$

where l_0 is the original length of the member and μ is its stiffness. The elastic support G is treated as a member with a tensile force $-R_4$, stiffness $0{\cdot}5\mu_0$ and original length zero, so that its extension is $-2R_4/\mu_0$.

Equation (4.7), in conjunction with equation (4.6), gives δl for each member in terms of known quantities and the four redundancies.

Compatibility

The correct values of the redundancies which ensure compatibility are again found by transforming the actual geometrical problem into four hypothetical equilibrium problems. Each of the force systems r_1, r_2, r_3 and r_4 satisfies all the requirements of equilibrium with zero external loads, except for loads at rigid supports in some cases. When used in the virtual work equation, the virtual work done is therefore zero in each case. Again, the actual bar extensions are used, since these are known to be compatible. It follows that

$$\sum_{\text{bars}} r_1(\lambda + e) = \sum_{\text{bars}} r_1[\alpha T l_0 + P/\mu] = 0 \tag{4.8}$$

together with three similar equations with r_2, r_3 and r_4 replacing r_1. These four equations then constitute the compatibility equations.

Using equation (4.6) for P, it follows from equation (4.8) that

$$\sum_{\text{bars}} r_1[\alpha T l_0 + (P_W + r_1 R_1 + r_2 R_2 + r_3 R_3 + r_4 R_4)/\mu] = 0$$

Expanding this equation, and writing also the other three equations with r_2, r_3 and r_4 substituted in turn for r_1, outside the square brackets, it is found that the four compatibility equations are as follows:

$$\left.\begin{aligned}
&R_1 \sum \frac{r_1^2}{\mu} + R_2 \sum \frac{r_1 r_2}{\mu} + R_3 \sum \frac{r_1 r_3}{\mu} \\
&\qquad + R_4 \sum \frac{r_1 r_4}{\mu} + \sum \frac{r_1 P_W}{\mu} + \alpha T \sum r_1 l_0 = 0 \\
&R_1 \sum \frac{r_2 r_1}{\mu} + R_2 \sum \frac{r_2^2}{\mu} + R_3 \sum \frac{r_2 r_3}{\mu} \\
&\qquad + R_4 \sum \frac{r_2 r_4}{\mu} + \sum \frac{r_2 P_W}{\mu} + \alpha T \sum r_2 l_0 = 0 \\
&R_1 \sum \frac{r_3 r_1}{\mu} + R_2 \sum \frac{r_3 r_2}{\mu} + R_3 \sum \frac{r_3^2}{\mu} \\
&\qquad + R_4 \sum \frac{r_3 r_4}{\mu} + \sum \frac{r_3 P_W}{\mu} + \alpha T \sum r_3 l_0 = 0 \\
&R_1 \sum \frac{r_4 r_1}{\mu} + R_2 \sum \frac{r_4 r_2}{\mu} + R_3 \sum \frac{r_4 r_3}{\mu} \\
&\qquad + R_4 \sum \frac{r_4^2}{\mu} + \sum \frac{r_4 P_W}{\mu} + \alpha T \sum r_4 l_0 = 0
\end{aligned}\right\} \quad (4.9)$$

In these equations each summation is taken over all the bars, together with the flexible support at G.

It will be noticed that the coefficients of the unknowns R_1, R_2, R_3 and R_4 in these four equations form a symmetrical matrix; thus there are ten summations involved in determining these coefficients and a further eight for the load and temperature terms. These summations are readily evaluated from the data in Table 4.4; details of this working will not be given. It is found that the compatibility equations are

$$4R_1+0{\cdot}5R_2+0{\cdot}5\sqrt{2}\,.\,R_3+1{\cdot}5\sqrt{2}\,.\,R_4 = 0$$

$$0{\cdot}5R_1+4R_2+0{\cdot}5\sqrt{2}\,.\,R_3+1{\cdot}5\sqrt{2}\,.\,R_4 = 2\sqrt{2}\,.\,W$$

$$0{\cdot}5\sqrt{2}\,.\,R_1+0{\cdot}5\sqrt{2}\,.\,R_2+4R_3+3R_4 = 6W+4\mu_0\alpha TL$$

$$1{\cdot}5\sqrt{2}\,.\,R_1+1{\cdot}5\sqrt{2}\,.\,R_2+3R_3+8{\cdot}5R_4 = 8W$$

The solution of these equations is

$$R_1 = -0{\cdot}5603W+0{\cdot}0165\mu_0\alpha TL$$

$$R_2 = 0{\cdot}2478W+0{\cdot}0165\mu_0\alpha TL$$

$$R_3 = 1{\cdot}0756W+1{\cdot}3605\mu_0\alpha TL$$

$$R_4 = 0{\cdot}6395W-0{\cdot}4884\mu_0\alpha TL$$

Knowing the values of the four redundancies, the forces in the bars are then readily determined from equation (4.6). The solution may then be checked by evaluating the extension and then verifying equation (4.8) directly.

The structural analysis cannot be said to be complete until the deflections of the joints are found. This question is discussed generally in Chapter 5, and the determination of deflections for this particular truss is given in Section 5.3.

The compatibility equations, such as equations (4.9) in this particular example, are often referred to as the flexibility coefficient equations, the coefficients of the redundancies forming the flexibility matrix. The procedure for deriving these equations for trusses was first established by Maxwell (1864) and independently by Mohr (1874), and so they are also termed the Maxwell–Mohr equations. A more general treatment for different types of structure was given by Müller–Breslau (1884).

Solution by Engesser's Theorem of Compatibility

Again it can be shown that precisely the same compatibility equations are established by the use of Engesser's Theorem of

Compatibility. The complementary energy c for a bar was defined by equation (3.18) as

$$c = P\lambda + \int_0^P e \, \mathrm{d}P$$

In this particular problem, $\lambda = \alpha T l_0$ and $e = P/\mu$, so that

$$c = P\alpha T l_0 + P^2/2\mu$$

Recalling that P is given as a linear function of the four redundancies by equation (4.6), namely

$$P = P_W + r_1 R_1 + r_2 R_2 + r_3 R_3 + r_4 R_4,$$

it follows that

$$\frac{\partial c}{\partial R_1} = \frac{\mathrm{d}c}{\mathrm{d}P}\frac{\partial P}{\partial R_1} = [\alpha T l_0 + P/\mu] r_1$$

Since the complementary energy C for the whole truss is obtained by summing c over all the bars, it follows at once that

$$\frac{\partial C}{\partial R_1} = \sum_{\text{bars}} r_1 [\alpha T l_0 + P/\mu]$$

Engesser's Theorem of Compatibility states that an equation of compatibility is obtained by setting $\partial C/\partial R_1 = 0$, which gives

$$\sum_{\text{bars}} r_1 [\alpha T l_0 + P/\mu] = 0$$

This is precisely equation (4.8), which was obtained by the virtual work transformation. The other three compatibility equations which are obtained by differentiating the complementary energy with respect to R_2, R_3 and R_4 are also easily shown to be identical with the last three of equations (4.9).

4.4. PROOF OF ENGESSER'S THEOREM OF COMPATIBILITY

A general proof of Engesser's Theorem of Compatibility for trusses is readily established by an extension of the above argument.

For any bar, the complementary energy c is given by

$$c = P\lambda + \int_0^P e\mathrm{d}P$$

If the truss has N redundancies, $R_1, R_2, \ldots, R_q, \ldots, R_n$, the force P in a typical bar can be expressed as

$$P = P_W + r_1R_1 + r_2R_2 + \ldots + r_qR_q + \ldots + r_nR_n$$

It follows that

$$\frac{\partial c}{\partial R_q} = \frac{\mathrm{d}c}{\mathrm{d}P}\frac{\partial P}{\partial R_q} = r_q\frac{\mathrm{d}c}{\mathrm{d}P}$$

In Chapter 3 it was shown (equation (3.20)) that

$$\frac{\mathrm{d}c}{\mathrm{d}P} = \lambda + e$$

Combining this with the above equation

$$\frac{\partial c}{\partial R_q} = r_q(\lambda + e),$$

and summing over all the bars, including any flexible supports as in the example of Section 4.3,

$$\frac{\partial C}{\partial R_q} = \sum_{\text{bars}} r_q(\lambda + e)$$

Thus Engesser's Theorem of Compatibility, which states that the compatibility equations are

$$\frac{\partial C}{\partial R_q} = 0, \qquad q = 1, 2, \ldots, N$$

is equivalent to the result

$$\sum_{\text{bars}} r_q(\lambda + e) = 0, \qquad q = 1, 2, \ldots, N$$

These equations are those which are derived immediately by the virtual work transformation, and the theorem is therefore established.

It should perhaps be noted that Engesser's Theorem of Compatibility is valid if the relationship between tension P and extension e caused by P for a bar is non-linear. It is, however, necessary that the relationship should be unique. Thus unloading of a bar, if the (P,e) relation then differs from the loading characteristic, must be excluded.

Minimum complementary energy

Engesser's Theorem of Compatibility states that the complementary energy C has a stationary value when the redundancies have the values which ensure compatibility. This stationary value can easily be shown to be a minimum for the special case in which λ is zero for every bar, and in addition the load/extension relation for each bar is the elastic law $P = \mu e$.

Let P be the actual force in any bar due to the actual external loads W, and let P^* be the force in the same bar in any hypothetical internal force system which satisfies all the requirements of equilibrium with the same external loads. It follows at once that the system of bar forces $(P-P^*)$ must satisfy all the requirements of equilibrium with zero external loads. Using this force system in conjunction with the actual bar extensions P/μ in the virtual work equation,

$$\sum_{\text{bars}} \frac{P}{\mu}(P-P^*) = 0 \tag{4.10}$$

Now consider the identity

$$\sum_{\text{bars}} \frac{1}{2\mu}(P^*-P)^2 \equiv \sum_{\text{bars}} \frac{1}{2\mu}\left[(P^*)^2-P^2+2P(P-P^*)\right]$$

By virtue of equation (4.10), this reduces to

$$\sum_{\text{bars}} \frac{1}{2\mu}(P^*-P)^2 = \sum_{\text{bars}} \frac{1}{2\mu}(P^*)^2-\sum_{\text{bars}} \frac{1}{2\mu}P^2$$

The left-hand side of this equation cannot be less than zero, and so

$$\sum_{\text{bars}} \frac{1}{2\mu} P^2 \leqslant \sum_{\text{bars}} \frac{1}{2\mu} (P^*)^2 \tag{4.11}$$

The left-hand side of this inequality represents the actual complementary energy C, while the right-hand side is the complementary energy C^* associated with any hypothetical set of bar forces P^* which satisfy the requirements of equilibrium with the actual external loads W. Thus

$$C \leqslant C^*,$$

which is the required result.

4.5. LACK OF FIT

Consider the truss illustrated in Fig. 4.4(a), which has two redundancies. It is supposed that the final two bars which are inserted during assembly are AE and CD, and that these bars are too long to fit between the joints by amounts λ_1 and λ_2, respectively. These bars would have to be inserted by jacking apart the joints A and E, and C and D by amounts equal to these lacks of fit, and the problem is to determine the forces in all the members after insertion of AE and CD and removal of the jacks.

There is little difficulty in dealing with such problems by the method of Sections 4.2 and 4.3. All that is necessary is to define clearly the datum geometry for the system, which is as depicted in Fig. 4.4(a), in which the distances between adjacent joints are all precisely L or $\sqrt{2} \cdot L$.

It will be supposed initially that there exists a unique, but not necessarily linear, relation $e = f(P)$ between the axial force P in each member and the extension e caused by this force.

Equilibrium

It is convenient to treat the bars AE and CD as the two redundant bars, the tensions in these bars being finally R_1 and R_2, respectively. The requirements of equilibrium are first analysed by considering the force systems illustrated in Figs. 4.4(b) and (c). In Fig. 4.4(b), the force in AE is unity whereas

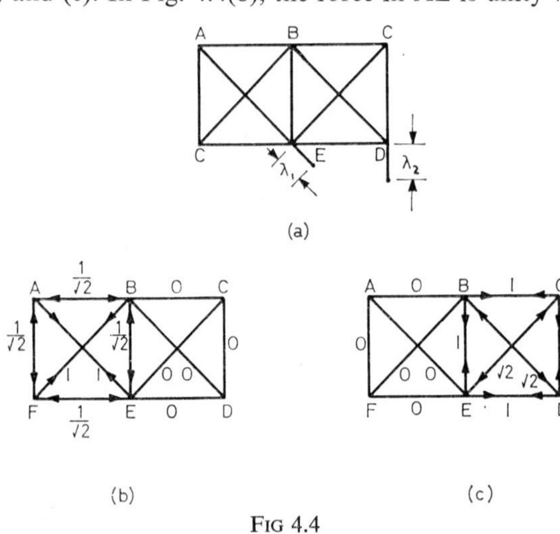

FIG 4.4

the force in CD is zero; the force in any bar in this system will be denoted by r_1. In Fig. 4.4(c) the force in AE is zero whereas the force in CD is unity, and the force in any bar in this system will be denoted by r_2. Values of r_1 and r_2 are given in Table 4.5. Since there are no external loads, it follows from the Principle of Superposition of forces that the actual force P in any bar is given by

$$P = r_1R_1+r_2R_2$$

Member characteristics

The extension e of any bar due to the force P is given by

$$e = f(P) = f(r_1R_1+r_2R_2)$$

The total extension δl of a bar from the datum length is defined as

$$\delta l = \lambda + e = \lambda + f(r_1 R_1 + r_2 R_2), \tag{4.12}$$

where λ is the increase in length due to all causes other than force. For this particular problem, λ is zero for every bar except AE and CD, and for these two bars it is equal to the lack of fit λ_1 or λ_2, respectively.

TABLE 4.5

Member	r_1	r_2	λ
AB	$-1/\sqrt{2}$	0	0
AF	$-1/\sqrt{2}$	0	0
FE	$-1/\sqrt{2}$	0	0
BF	1	0	0
AE	1	0	λ_1
BE	$1/\sqrt{2}$	1	0
BC	0	1	0
CD	0	1	λ_2
DE	0	1	0
BD	0	$-\sqrt{2}$	0
CE	0	$-\sqrt{2}$	0

Compatibility

The extensions δl must satisfy the requirements of compatibility, since they are the changes in length of the bars from the original datum to the final situation, and can therefore be used in the equation of virtual work. Two force systems are used in conjunction with these extensions, these being the systems r_1 and r_2. There are no external loads in either force system, so that the virtual work done by external loads is zero. It follows that

$$\sum_{\text{bars}} r_1[\lambda + f(r_1 R_1 + r_2 R_2)] = 0$$

$$\sum_{\text{bars}} r_2[\lambda + f(r_1 R_1 + r_2 R_2)] = 0$$

These are the two equations of compatibility from which R_1 and R_2 can be determined.

To illustrate the process, suppose that all the bars are elastic and have the same stiffness μ_0, so that $e = P/\mu_0$. Then

$$e = f(r_1R_1+r_2R_2) = (r_1R_1+r_2R_2)/\mu_0$$

and the compatibility equations become

$$R_1 \sum r_1^2+R_2 \sum r_1r_2+\mu_0 \sum r_1\lambda = 0 \tag{4.13}$$

$$R_1 \sum r_2r_1+R_2 \sum r_2^2+\mu_0 \sum r_2\lambda = 0 \tag{4.14}$$

The summations involved in equations (4.13) and (4.14) are readily determined from the data in Table 4.5, and the compatibility equations are found to be

$$4R_1+\frac{1}{\sqrt{2}}R_2+\mu_0\lambda_1 = 0$$

$$\frac{1}{\sqrt{2}}R_1+8R_2+\mu_0\lambda_2 = 0$$

Their solution is

$$R_1 = \mu_0\left[\frac{\sqrt{2}.\lambda_2-16\lambda_1}{63}\right]$$

$$R_2 = \mu_0\left[\frac{\sqrt{2}.\lambda_1-8\lambda_2}{63}\right]$$

The forces in the bars can then be found from the data in Table 4.5.

4.6. CASTIGLIANO'S THEOREM OF COMPATIBILITY

The problem just described could have been solved by applying Engesser's Theorem of Compatibility; it has already been shown in Section 4.4 that this theorem achieves precisely the same transformation as the virtual work approach adopted. It could also have been solved using Castigliano's Theorem of

Compatibility (Castigliano, 1879); this may be regarded as a special form of Engesser's Theorem of Compatibility when the following two conditions are fulfilled.

(i) λ is zero for all the bars of the basic truss, and is the lack of fit for each redundant bar.
(ii) The (P,e) relation for each bar is linear elastic.

Castigliano's Theorem of Compatibility can be stated as follows. Let a truss have N redundancies, namely $R_1, R_2, \ldots, R_q, \ldots, R_n$. The N redundant bars are inserted with lacks of fit $\lambda_1, \lambda_2, \ldots, \lambda_q, \ldots, \lambda_n$. Then if U denotes the strain energy for the whole truss, the N compatibility equations are:

$$\frac{\partial U}{\partial R_q}+\lambda_q = 0, \qquad q = 1, 2, \ldots, N \tag{4.15}$$

To show that this result is a special case of Engesser's Theorem, it is noted that for each bar, the complementary energy c is defined as

$$c = P\lambda+\int_0^P e\,\mathrm{d}P,$$

so that if

$$P = \mu e \qquad \text{(Hooke's Law)},$$

$$c = P\lambda+P^2/2\mu$$

The strain energy u for each bar is defined as

$$u = \int_0^e P\,\mathrm{d}e = \tfrac{1}{2}\mu e^2 = P^2/2\mu,$$

using Hooke's Law. It follows that

$$c = P\lambda+u$$

The force P in any bar is given in terms of the redundancies as

$$P = P_W+r_1R_1+r_2R_2+\ldots+r_qR_q+\ldots+r_nR_n$$

It follows that

$$\frac{\partial c}{\partial R_q} = \frac{\partial P}{\partial R_q}\lambda+\frac{\partial u}{\partial R_q} = r_q\lambda+\frac{\partial u}{\partial R_q},$$

and summing over all the bars of the truss,

$$\frac{\partial C}{\partial R_q} = \sum_{\text{bars}} r_q \lambda + \frac{\partial U}{\partial R_q}$$

Now λ is only non-zero for the redundant bars, so that the summation in this equation need only be extended over these bars. Further, for the redundant bar q, $r_q = 1$, whereas for all the other redundant bars, $r_q = 0$. The above equation therefore reduces to

$$\frac{\partial C}{\partial R_q} = \lambda_q + \frac{\partial U}{\partial R_q}$$

Engesser's Theorem of Compatibility, which has already been proved, states that $\partial C/\partial R_q = 0$; it follows that an equivalent statement under the restrictive conditions (i) and (ii) is

$$\frac{\partial U}{\partial R_q} + \lambda_q = 0,$$

and this is Castigliano's Theorem of Compatibility.

4.7. FURTHER APPLICATIONS OF VIRTUAL WORK

So far the discussion has been concerned solely with trusses. Some other simple types of structure, namely a portal frame, an arch and a portion of a ring loaded normal to its plane, will now be considered. In each case the compatibility equations will be determined by the virtual work method. The application of the appropriate energy theorem will also be described briefly.

Portal frame

The rigidly jointed portal frame whose dimensions and loading are as shown in Fig. 4.5 is composed of uniform members which behave elastically and have flexural rigidity EI. It is required to determine the bending moment distribution throughout the frame due to the loading, the frame being free from stress when unloaded.

This problem has two redundancies, because of the symmetry of the frame and loading. These may be taken as the bending moments m_1 at A and D, and m_2 at B and C. Initially, the values of m_1 and m_2 which ensure compatibility are sought.

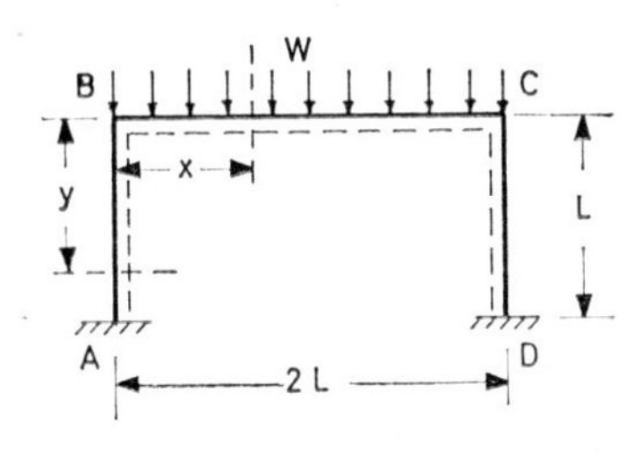

FIG 4.5

Equilibrium

As is customary in frame analyses, it is assumed that deformations due to shear and direct thrust are negligible. Thus the first step in the analysis, which in general is to determine the distribution of internal forces throughout the structure in terms of the redundancies, involves expressing the bending moment M at every section in terms of m_1 and m_2.

The sign convention which will be adopted for bending moment is that a positive moment causes tension in those fibres of the member adjacent to the dotted line in Fig. 4.5.

The statical analysis may conveniently be carried out by expressing M in the form

$$M = M_W + \alpha_1 m_1 + \alpha_2 m_2 \tag{4.16}$$

M_W is the bending moment at any section due to the loading with m_1 and m_2 both zero. α_1 represents the bending moment which would occur at any section if the external load was zero and m_2 was also zero, with $m_1 = 1$. Similarly, α_2 is the bending moment which would occur with zero external load and zero value of m_1, with $m_2 = 1$. Equation (4.16) then merely expresses the Principle of Superposition of forces. The coefficients α_1 and α_2 are, of course, functions of position in the frame.

Values of M_W, α_1 and α_2 are given in Table 4.6. Because of symmetry, only one half of the frame is considered in the analysis, and so Table 4.6 only deals with members AB and BC (up to its mid-point).

TABLE 4.6

Member	M_W	α_1	α_2
AB	0	$\left(\frac{y}{L}\right)$	$1-\left(\frac{y}{L}\right)$
BC	$WL\left(\frac{x}{2L}\right)\left[1-\left(\frac{x}{2L}\right)\right]$	0	1

Member characteristics

The deformations to be considered are the changes of curvature due to bending. These are given by

$$\kappa = \frac{M}{EI} = \frac{1}{EI}[M_W + \alpha_1 m_1 + \alpha_2 m_2] \tag{4.17}$$

Compatibility

The Principle of Virtual Work for frames was given in equation (3.6) as

$$\sum_{\text{joints}} (H^*h^{**} + V^*v^{**} + C^*\phi^{**}) + \oint (w_n^* y_n^{**} + w_t^* y_t^{**})\, ds$$
$$= \sum_{\text{hinges}} M^*\theta^{**} + \oint (M^*\kappa^{**} + P^*\varepsilon^{**})\, ds$$

As usual in the compatibility method, the actual deformations are used in the virtual work equation. Two force systems are used in turn, these being the bending moment distributions α_1 and α_2. Each of these distributions satisfies the requirements of equilibrium with zero external loads except at the feet A and D, where the displacements are zero, so that all the terms on the left-hand side of the virtual work equation become zero. Because the knee joints B and C are rigid, the term $\Sigma M^*\theta^{**}$ is also zero. Further, the axial strains ε in the frame are presumed to have negligible effects as compared with the effects of curvature, so that the term involving $P^*\varepsilon^{**}$ is also zero. The two virtual work equations therefore reduce to:

$$\oint \alpha_1 \kappa \, ds = 0, \qquad \oint \alpha_2 \kappa \, ds = 0 \tag{4.18}$$

Substituting the value of κ given by equation (4.17), equations (4.18) become

$$\oint \alpha_1 \frac{1}{EI} [M_W + \alpha_1 m_1 + \alpha_2 m_2] \, ds = 0 \tag{4.19}$$

$$\oint \alpha_2 \frac{1}{EI} [M_W + \alpha_1 m_1 + \alpha_2 m_2] \, ds = 0 \tag{4.20}$$

Equations (4.19) and (4.20) are the two equations of compatibility; since EI is constant throughout the frame they reduce to:

$$m_1 \oint \alpha_1^2 \, ds + m_2 \oint \alpha_1 \alpha_2 \, ds + \oint \alpha_1 M_W \, ds = 0 \tag{4.21}$$

$$m_1 \oint \alpha_2 \alpha_1 \, ds + m_2 \oint \alpha_2^2 \, ds + \oint \alpha_2 M_W \, ds = 0 \tag{4.22}$$

The integrals involved in these equations are readily evaluated from the data in Table 4.6, and it is found that

$$m_1(\tfrac{1}{3}L) + m_2(\tfrac{1}{6}L) = 0$$

$$m_1(\tfrac{1}{6}L) + m_2(\tfrac{4}{3}L) + \tfrac{1}{6}WL^2 = 0$$

The solution of these equations is

$$m_1 = WL/15, \qquad m_2 = -2WL/15$$

The bending moment distribution throughout the frame is then readily determined with the aid of Table 4.6.

Solution by Castigliano's Theorem of Compatibility

This problem could also have been solved by Castigliano's Theorem of Compatibility. Since there is no lack of fit, this theorem (equation (4.15)) states that the compatibility equations are

$$\frac{\partial U}{\partial R_q} = 0, \qquad q = 1, 2, \ldots, N,$$

where $R_1, R_2, \ldots, R_q, \ldots, R_n$ are the N redundancies. In this particular case there are two redundancies, m_1 and m_2, so that the two compatibility equations are

$$\frac{\partial U}{\partial m_1} = 0, \qquad \frac{\partial U}{\partial m_2} = 0$$

For an element of beam of length δs, the strain energy δu is given by equation (3.15) as

$$\frac{du}{ds} = \int_0^{\kappa} M \,d\kappa = \int_0^{\kappa} EI\kappa \,d\kappa = \tfrac{1}{2}EI\kappa^2 = \frac{M^2}{2EI},$$

so that for a whole frame (or in this case one half of the frame)

$$U = \oint \frac{M^2}{2EI} \,ds$$

Using equation (4.16),

$$U = \oint \frac{(M_W + \alpha_1 m_1 + \alpha_2 m_2)^2}{2EI} \,ds,$$

so that $$\frac{\partial U}{\partial m_1} = \oint \alpha_1 \frac{1}{EI}[M_W + \alpha_1 m_1 + \alpha_2 m_2] \,ds$$

According to Castigliano's Theorem of Compatibility, $\partial U/\partial m_1 = 0$, so that

$$\oint \alpha_1 \frac{1}{EI}[M_W + \alpha_1 m_1 + \alpha_2 m_2] \,ds = 0$$

This equation is identical with equation (4.19), the first of the two equations of compatibility established by the virtual work transformation, and it is easily shown that the equation $\partial U/\partial m_2 = 0$ is identical with equation (4.20). Thus Castigliano's Theorem of Compatibility leads to precisely the same compatibility equations as were obtained by the virtual work method.

Temperature stresses in arch

The arch illustrated in Fig. 4.6 is pinned to a rigid abutment at O and rigidly built in at A. The profile is parabolic, with a rise h, the equation of the centre line being

$$z = \frac{4h}{L^2}x(L-x) \tag{4.23}$$

The arch is presumed to behave elastically, and the flexural rigidity of the section is given by

$$EI = EI_0 \sec \psi, \tag{4.24}$$

where EI_0 is the flexural rigidity at the crown of the arch. It is required to find the maximum bending moment induced by a

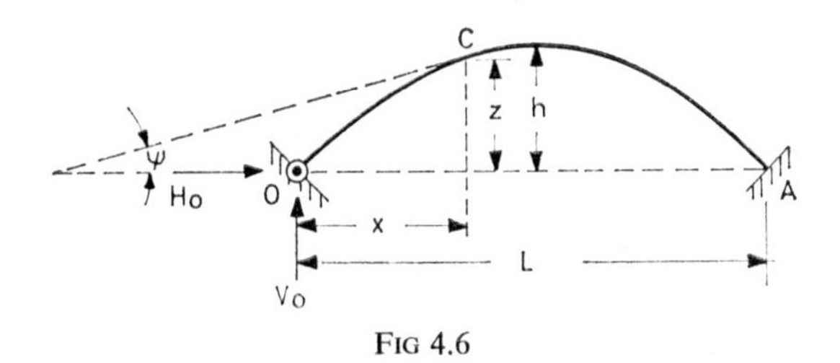

FIG 4.6

temperature rise T above a datum temperature at which the arch is free from stress, the coefficient of linear expansion being α.

The arch has two redundancies, which may be specified as the horizontal and vertical components of reaction, H_0 and V_0

respectively, at O. These will be determined from two equations of compatibility, which will be established using the virtual work method.

As in the previous problem, the deformations caused by axial thrust and by shear force will be presumed to have negligible effects as compared with the deformations due to bending.

The Principle of Virtual Work in the form quoted for frames, namely

$$\sum_{\text{joints}} (H^*h^{**}+V^*v^{**}+C^*\phi^{**})+\oint (w_n^*y_n^{**}+w_t^*y_t^{**})\,ds = \sum_{\text{joints}} M^*\theta^{**}+\oint (M^*\kappa^{**}+P^*\varepsilon^{**})\,ds$$

is equally applicable to this problem. Again the actual deformations of the arch will be used, and the two equations of compatibility will be established by using the force systems corresponding to $H_0 = 1$, $V_0 = 0$ and $H_0 = 0$, $V_0 = 1$.

In each of these force systems there are no external loads on the arch rib. There are no displacements of any kind at A, and only a rotation is possible at O, where there is no moment at the pin-joint. The virtual work equation therefore reduces to:

$$\oint (M^*\kappa^{**}+P^*\varepsilon^{**})\,ds = 0 \tag{4.25}$$

By contrast with the previous problem, the term $P^*\varepsilon^{**}$ cannot be ignored. The axial strain due to axial force in the arch rib is still presumed to be negligible, but the axial strain due to the temperature rise T is not negligible, and indeed is the cause of the bending of the arch.

Equilibrium

The first step in the analysis is to express the bending moment and axial force at any section in terms of the redundancies H_0 and V_0. Defining M as positive when hogging and P as positive when tensile, it is seen from Fig. 4.6 that

$$M = H_0z-V_0x \tag{4.26}$$

$$P = -H_0\cos\psi-V_0\sin\psi \tag{4.27}$$

Member characteristics

The actual hogging curvature κ at any section is given by

$$\kappa = \frac{M}{EI} = \frac{1}{EI}[H_0 z - V_0 x], \tag{4.28}$$

and the actual axial tensile strain ε is

$$\varepsilon = \alpha T \tag{4.29}$$

Compatibility

These actual deformations κ and ε are used in the equation of virtual work, equation (4.25). Two force systems are used in turn; the first is obtained by setting $H_0 = 1$ and $V_0 = 0$ in equations (4.26) and (4.27), and the second corresponds to $H_0 = 0$, $V_0 = 1$. The equations of compatibility thus obtained are

$$\int_0^A z \frac{1}{EI}[H_0 z - V_0 x]\, ds + \int_0^A -\cos\psi(\alpha T)\, ds = 0 \tag{4.30}$$

$$\int_0^A -x \frac{1}{EI}[H_0 z - V_0 x]\, ds + \int_0^A -\sin\psi(\alpha T)\, ds = 0 \tag{4.31}$$

These equations may be simplified by noting that $ds \cos\psi = dx$, $ds \sin\psi = dz$, and also from equation (4.24)

$$\frac{ds}{EI} = \frac{ds}{EI_0 \sec\psi} = \frac{dx}{EI_0}$$

Using these results, equations (4.30) and (4.31) become

$$H_0 \int_0^A z^2 \frac{dx}{EI_0} - V_0 \int_0^A zx \frac{dx}{EI_0} - \alpha T \int_0^A dx = 0$$

$$-H_0 \int_0^A xz \frac{dx}{EI_0} + V_0 \int_0^A x^2 \frac{dx}{EI_0} - \alpha T \int_0^A dz = 0$$

Evaluating the integrals by making use of equation (4.23),

$$\tfrac{8}{15} H_0 h^2 L - \tfrac{1}{3} V_0 h L^2 = EI_0 \alpha T L$$

$$-\tfrac{1}{3} H_0 h L^2 + \tfrac{1}{3} V_0 L^3 = 0$$

The solution of these equations is

$$H_0 = 5EI_0\alpha T/h^2$$
$$V_0 = 5EI_0\alpha T/hL$$

The hogging bending moment M at any section is therefore

$$M = H_0 z - V_0 x = \frac{5EI_0\alpha T}{h}\left[3\left(\frac{x}{L}\right)-4\left(\frac{x}{L}\right)^2\right],$$

and from this expression it can be shown that the largest bending moment occurs at A, where the sagging moment is $5EI_0\alpha T/h$.

Solution by Engesser's Theorem of Compatibility

Engesser's Theorem of Compatibility may also be used to solve this problem. The complementary energy must be considered in two parts, one part being associated with flexure and the other with axial strain.

As far as flexure is concerned, the curvature change κ is due entirely to bending moment. The complementary energy δc in a length δs due to flexure is therefore given by equation (3.22) as

$$\frac{\mathrm{d}c}{\mathrm{d}s} = \int_0^M \kappa\,\mathrm{d}M = \int_0^M \frac{M}{EI}\,\mathrm{d}M = \frac{M^2}{2EI} \tag{4.32}$$

The complementary energy associated with extension has been defined by equation (3.18) as

$$c = P\lambda + \int_0^P e\,\mathrm{d}P,$$

where e is the extension due to P and λ is the change of length due to any other causes. If the extension due to axial load may be neglected, $c = P\lambda$, and so for an element of the arch of length δs, for which $\lambda = \alpha T\delta s$,

$$\frac{\mathrm{d}c}{ds} = P\alpha T \tag{4.33}$$

Adding together the contributions to dc/ds from equations (4.32) and (4.33),

$$\frac{dc}{ds} = \frac{M^2}{2EI} + P\alpha T,$$

so that for the whole arch,

$$C = \int_0^A \left[\frac{M^2}{2EI} + P\alpha T\right] ds$$

Inserting the values of M and P given in equations (4.26) and (4.27),

$$C = \int_0^A \left[\frac{(H_0 z - V_0 x)^2}{2EI} + (-H_0 \cos\psi - V_0 \sin\psi)\alpha T\right] ds \quad (4.34)$$

Engesser's Theorem of Compatibility states that the equations of compatibility are given by

$$\frac{\partial C}{\partial H_0} = 0, \qquad \frac{\partial C}{\partial V_0} = 0$$

It will be seen from equation (4.34) that these equations are identical with equations (4.30) and (4.31), which were derived by virtual work.

Encastré semi-circular ring loaded normal to its plane

The last problem to be considered involves torsion as well as bending. Figures 4.7(a) and (b) show a rod of circular cross-section which is bent into the form of a semi-circle of radius R lying in a horizontal plane. The rod is rigidly built in at A and B, and carries a vertical concentrated load W at its mid-point, as shown. It is required to find the bending moment and torque acting at any cross-section, the rod being free from stress when unloaded. The rod is assumed to behave elastically; the flexural rigidity of the section is EI and the torsional rigidity is GJ.

By considerations of statics and of symmetry it can be seen

that the vertical reactions at A and B are each $W/2$. Moreover, there must be supporting bending moments $WR/2$ at each of these two sections, as shown in the figures. The double arrow-head notation used for their representation implies that the

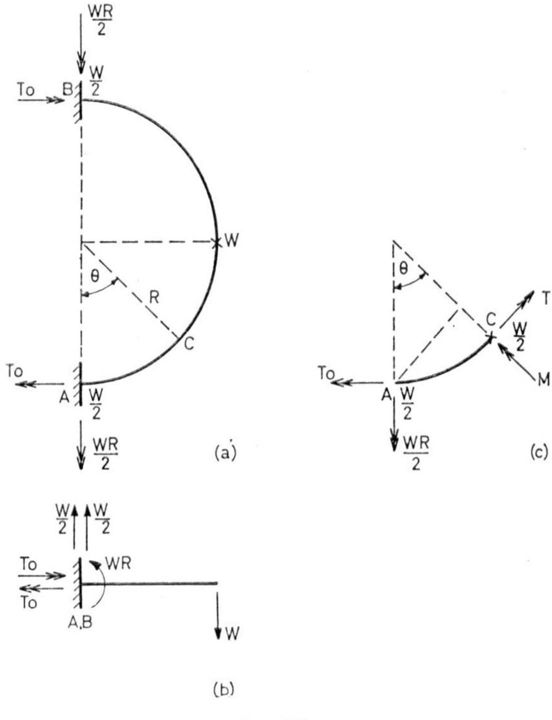

FIG 4.7

sense of a positive couple corresponds to a right-hand screw motion when looking in the direction of these arrows.

The torques T_0 which act at A and B cannot be determined by statics or symmetry, and so T_0 is the single redundancy of the problem, and must be found by establishing an equation of compatibility.

Equilibrium

Figure 4.7(c) shows a plan view of a segment AC of the ring which subtends an angle θ at the centre O. Bearing in mind that there is an upward reaction $W/2$ at A and a corresponding downwards shear force $W/2$ at C, it is seen that the bending moment M and torque T at C are given by

$$M = \frac{WR}{2}\cos\theta - T_0 \sin\theta - \frac{WR}{2}\sin\theta \tag{4.35}$$

$$T = \frac{WR}{2}\sin\theta + T_0 \cos\theta - \frac{WR}{2}(1-\cos\theta), \tag{4.36}$$

these expressions being valid for values of θ between 0 and $\pi/2$.

Member characteristics

The actual curvature κ at any section is M/EI and the angle of twist per unit length τ is T/GJ. These deformations are known to satisfy the requirements of compatibility and are used in the equation of virtual work.

Compatibility

In order to establish the compatibility equation, the system of internal forces corresponding to unit value of the redundancy T_0 and with zero external load is used. This system is seen from equations (4.35) and (4.36) to be

$$M = -\sin\theta$$

$$T = \cos\theta$$

Since the rod is rigidly built in at A and B, there are no rotations corresponding to T_0, and since the external load is zero the virtual work done on the ring is zero. The virtual work absorbed, which is associated with the changes of curvature and twist per unit length, is therefore zero, so that

$$\int_0^{\pi/2} -\sin\theta(\kappa)\,ds + \int_0^{\pi/2} \cos\theta(\tau)\,ds = 0$$

It is of course only necessary to integrate over one half of the ring because of symmetry.

Substituting $\kappa = M/EI$, $\tau = T/GJ$, and using equations (4.35) and (4.36), the compatibility equation becomes

$$\frac{1}{EI}\int_0^{\pi/2} -\sin\theta\left[\frac{WR}{2}\cos\theta - T_0\sin\theta - \frac{WR}{2}\sin\theta\right]Rd\theta$$

$$+\frac{1}{GJ}\int_0^{\pi/2}\cos\theta\left[\frac{WR}{2}\sin\theta + T_0\cos\theta - \frac{WR}{2}(1-\cos\theta)\right]Rd\theta = 0 \quad (4.37)$$

Evaluating the integrals it is found that

$$\left(\frac{1}{EI}+\frac{1}{GJ}\right)\left[\frac{\pi}{4}T_0 + \frac{WR}{2}\left(\frac{\pi}{4}-\frac{1}{2}\right)\right] = 0,$$

so that

$$T_0 = -WR\left(\frac{1}{2}-\frac{1}{\pi}\right)$$

Using equations (4.35) and (4.36) it then follows that

$$M = \frac{WR}{2}\left(\cos\theta - \frac{2}{\pi}\sin\theta\right),$$

$$T = \frac{WR}{2}\left(\sin\theta + \frac{2}{\pi}\cos\theta - 1\right), \qquad 0 \leqslant \theta \leqslant \frac{\pi}{2}$$

Solution by Castigliano's Theorem of Compatibility

The compatibility condition, equation (4.37), can also be obtained by applying Castigliano's Theorem of Compatibility. The strain energy U for one half of the ring is given by

$$U = \int_0^{\pi/2}\frac{M^2}{2EI}\,ds + \int_0^{\pi/2}\frac{T^2}{2GJ}\,ds$$

and using equations (4.35) and (4.36) this becomes

$$U = \frac{1}{2EI}\int_0^{\pi/2}\left[\frac{WR}{2}\cos\theta - T_0\sin\theta - \frac{WR}{2}\sin\theta\right]^2 Rd\theta$$

$$+\frac{1}{2GJ}\int_0^{\pi/2}\left[\frac{WR}{2}\sin\theta + T_0\cos\theta - \frac{WR}{2}(1-\cos\theta)\right]^2 Rd\theta$$

Castigliano's Theorem of Compatibility states that the correct value of T_0 which ensures compatibility is furnished by the equation

$$\frac{\partial U}{\partial T_0} = 0,$$

and it is seen at once that this yields precisely equation (4.37).

5

Calculation of deflections

5.1. INTRODUCTION

This chapter will be concerned solely with the problem of determining the deflections of a structure when the deformations of its individual members are known. As already pointed out in Chapter 3, a direct geometrical approach is always possible, but can often be tedious, so that a transformation to a hypothetical equilibrium problem by using the Principle of Virtual Work is usually advantageous.

The virtual work transformation, or "dummy unit load" method as it is often called, was referred to briefly in Section 3.3. In the present chapter it will first be considered in detail in relation to statically determinate and indeterminate trusses. It will also be shown that precisely equivalent results are obtained by applying the following energy theorems in appropriate cases:

First Theorem of Complementary Energy, and
Castigliano's Theorem (Part II).

The latter theorem is shown to be a special case of the First Theorem of Complementary Energy for structures obeying Hooke's Law and in which the deflections are due solely to the application of loads. Finally, consideration is given to the determination of the deflections of beams, including the problem of determining deflections at the point of plastic collapse.

5.2. STATICALLY DETERMINATE TRUSS

The simple truss illustrated in Fig. 5.1 will first be considered as an example. In this truss all the bars have a coefficient of linear expansion α and are of original length L or $\sqrt{2}\,.\,L$, and all the angles between bars are 45° or 90°. It is required to

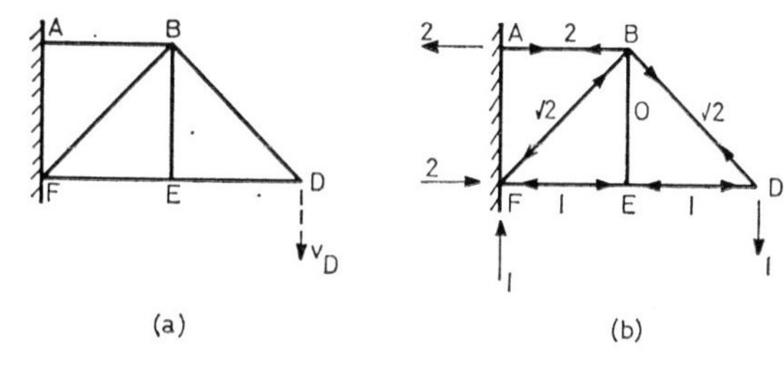

FIG 5.1

determine the vertical deflection v_D of D caused by a temperature rise T.

Since there are no external loads and the truss is statically determinate, no forces are induced in the bars by the temperature rise. The extension δl of each bar is defined as

$$\delta l = \lambda + e,$$

where e is the extension due to axial force, and λ is the extension due to all other causes. In this particular case, e is zero, so that

$$\delta l = \lambda = \alpha T l_0.$$

Since the extensions λ are actual extensions, they must be compatible with the actual joint deflections, and in particular with v_D, so that they can be used in the virtual work equation. Following the method of Section 3.3, a hypothetical force system is used which consists of the bar forces P' satisfying the conditions of equilibrium with a unit vertical load at D. Since the truss is statically determinate, these forces are readily determined; the complete force system is depicted in Fig. 5.1(b).

The equation of virtual work, equation (3.1a), is as follows:

$$\sum_{\text{joints}} (H^*h^{**} + V^*v^{**}) = \sum_{\text{bars}} P^*(\delta l)^{**}$$

The hypothetical force system just described is used in this equation in conjunction with the actual bar extensions and joint deflections. Considering first the left-hand side of the virtual work equation, it is seen that at A and F, h^{**} and v^{**} are zero, and at B and E, H^* and V^* are both zero. The only term arising is in fact $1 \times v_D$ at joint D. It follows that

$$v_D = \sum_{\text{bars}} P'\delta l = \sum_{\text{bars}} P'\lambda \tag{5.1}$$

Values of l_0, λ, P' and $P'\lambda$ are given in Table 5.1.

TABLE 5.1

Member	AB	BD	DE	EF	BE	BF
l_0	L	$\sqrt{2}.L$	L	L	L	$\sqrt{2}.L$
λ	αTL	$\sqrt{2}.\alpha TL$	αTL	αTL	αTL	$\sqrt{2}.\alpha TL$
P'	2	$\sqrt{2}$	-1	-1	0	$-\sqrt{2}$
$P'\lambda$	$2\alpha TL$	$2\alpha TL$	$-\alpha TL$	$-\alpha TL$	0	$-2\alpha TL$

From this table it is seen that $\Sigma P'\delta l = 0$, so that from equation (5.1), $v_D = 0$.

This is a characteristic example of the way in which the actual geometrical problem of determining a deflection can be transformed into a hypothetical equilibrium problem. The task of determining deflections by a geometrical method, which for a truss would involve the construction of a Williot diagram, is replaced by the problem of determining the forces in equilibrium with a dummy or hypothetical unit load imagined to be applied at the point where the deflection is required and in the direction of that deflection.

This procedure for the determination of deflections was first given by Maxwell (1864) for pin-jointed trusses. Maxwell's paper, which was reproduced and commented on by Niles (1950), went largely unnoticed, and the method was rediscovered and extended by Mohr (1874). The procedure may therefore be referred to as the Maxwell–Mohr method.

Solution by First Theorem of Complementary Energy

The above calculations could also have been based on the First Theorem of Complementary Energy (Engesser, 1889). In general, this theorem states that the total deflection y due to all causes is given by

$$y = \frac{\partial C}{\partial W} \tag{5.2}$$

In this expression y is the component of deflection at the point of application of a load W and in the direction of W.

In applying this theorem to the above problem the procedure is to calculate the vertical deflection of D when a vertical load W is applied at this joint and the temperature rise T has also taken place. An expression for v_D then results which is a function of W, and the required value of v_D is obtained by setting $W = 0$.

With a vertical load W applied at D, the force P in a typical bar would be WP'. The complementary energy c in a bar was defined by equation (3.18) as

$$c = P\lambda + \int_0^P e \, dP$$

It was shown in Chapter 3 (equation (3.20)) that as a consequence of this definition,

$$\frac{dc}{dP} = \lambda + e,$$

so that $$\frac{\partial c}{\partial W} = \left(\frac{dc}{dP}\right)\frac{\partial P}{\partial W} = (\lambda + e)P',$$

since $$P = WP'$$

Summing over the bars,

$$\frac{\partial C}{\partial W} = \sum_{\text{bars}} P'(\lambda + e),$$

so that from equation (5.2), expressing the First Theorem of Complementary Energy,

$$v_D = \sum_{\text{bars}} P'(\lambda + e) \tag{5.3}$$

Equation (5.3) expresses v_D as a function of W, since for each bar e is the extension due to the force P in the bar, which in this particular case is equal to WP'. Although the relationship between P and e is not known, e must be zero for each bar when W, and therefore P is zero. Thus for the actual problem being considered, equation (5.3) reduces to

$$v_D = \sum_{\text{bars}} P'\lambda,$$

and this is identical with equation (5.1), which was derived by virtual work.

5.3. STATICALLY INDETERMINATE TRUSS

The example just considered was statically determinate. The more general case of statically indeterminate structures will now be discussed, with particular reference to the truss shown in Fig. 5.2(a), which has four redundancies. All the bars in this truss are of the same stiffness μ_0, and the support at G has stiffness $0{\cdot}5\mu_0$; the loading is as shown and in addition there is a temperature rise T above a datum temperature at which the truss is free from stress when unloaded. It will be supposed that all the bar elongations δl due to the loading and the temperature rise are known, the redundancies having been determined by the compatibility method. (The solution for the redundancies was in fact given in Section 4.3.) It is required to determine the vertical deflection v_F of F.

As before, the actual bar elongations and joint deflections, which are known to satisfy the requirements of compatibility, are used in the virtual work equation. A system of bar forces

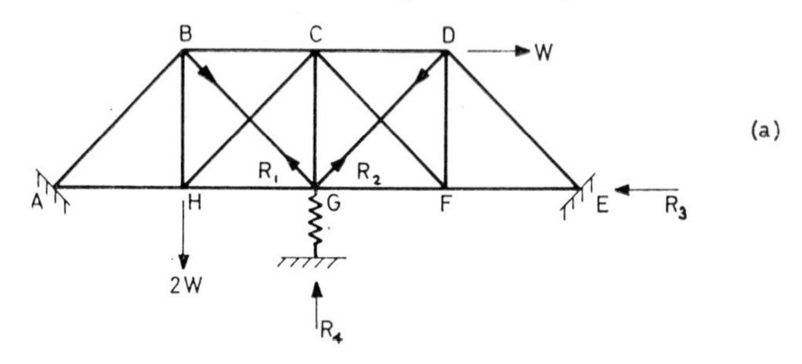

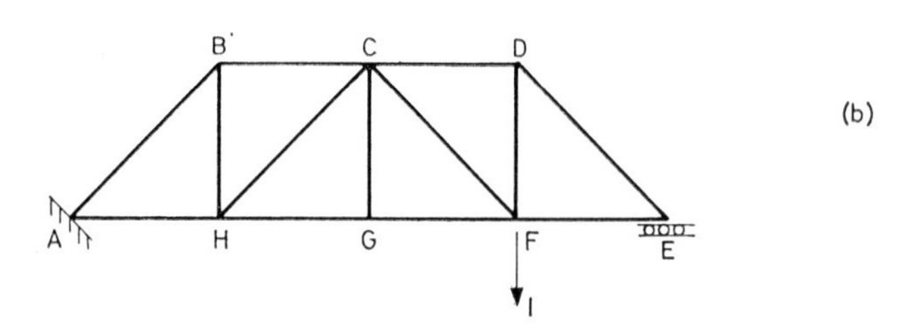

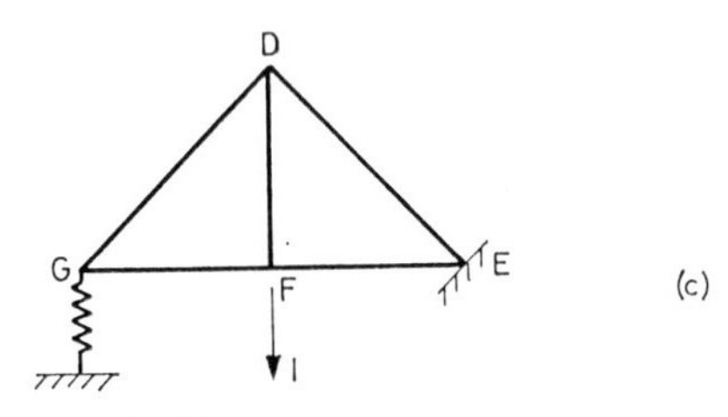

FIG 5.2

P' satisfying all the requirements of equilibrium with a hypothetical unit vertical load at F is also used, and it then follows that

$$v_F = \sum_{\text{bars}} P'\delta l, \tag{5.4}$$

the summation also including the spring support at G, which is merely treated as a bar of stiffness $0 \cdot 5\mu_0$.

This relation is true if the bar forces P' represent *any* system satisfying the requirements of equilibrium with the dummy unit load at F. In particular, it is not necessary to determine the forces which would occur in the bars if a unit load was actually applied at F; the determination of these forces would involve the solution of a problem with four redundancies. One procedure is to set the four redundancies equal to zero, so that the truss becomes statically determinate, as in Fig. 5.2(b). The system of forces P' in the remaining bars can then be determined by statics.

These forces are given in Table 5.2, in which the actual bar extensions δl are also tabulated from the data given in Table 4.4. Performing the summation of equation (5.4), it is then found that

$$v_F = \frac{1}{\mu_0}[4 \cdot 625W - 0 \cdot 5\sqrt{2} \, . \, R_1 - 2R_3 - 3 \cdot 5R_4] \quad (5.5)$$

The values of the redundancies found in Chapter 4 were:

$$R_1 = -0 \cdot 5603W + 0 \cdot 0165\mu_0 \alpha TL$$
$$R_2 = 0 \cdot 2478W + 0 \cdot 0165\mu_0 \alpha TL$$
$$R_3 = 1 \cdot 0756W + 1 \cdot 3605\mu_0 \alpha TL$$
$$R_4 = 0 \cdot 6395W - 0 \cdot 4884\mu_0 \alpha TL$$

Substituting in equation (5.5) it is then found that

$$v_F = 0 \cdot 6316W/\mu_0 - 1 \cdot 0233\alpha TL$$

Other force systems can be used for the determination of this deflection. It is only necessary to realise that, in general, the knowledge of the extensions of any statically determinate assemblage of bars suffices to locate the joints involved (Charlton, 1955). Thus in Fig. 5.2(b), the bars BG and DG and the spring support at G need not be considered since their extensions are known to be compatible with the distortions of the particular basic truss depicted; moreover, the extensions of the bottom chord members AH, HG, GF, and FE must add up to zero for compatibility with zero horizontal movement of E.

TABLE 5.2

Member	δl	P'	P''
AB	$\sqrt{2}.\alpha TL+\frac{1}{\mu_0}[-1\cdot25\sqrt{2}.W+0\cdot5\sqrt{2}.R_4]$	$-0\cdot25\sqrt{2}$	
BC	$\alpha TL+\frac{1}{\mu_0}[-1\cdot25W-0\cdot5\sqrt{2}.R_1+0\cdot5R_4]$	$-0\cdot25$	
CD	$\alpha TL+\frac{1}{\mu_0}[0\cdot25W-0\cdot5\sqrt{2}.R_2+0\cdot5R_4]$	$-0\cdot75$	
DE	$\sqrt{2}.\alpha TL+\frac{1}{\mu_0}[-0\cdot75\sqrt{2}.W+0\cdot5\sqrt{2}.R_4]$	$-0\cdot75\sqrt{2}$	$-0\cdot5\sqrt{2}$
AH	$\alpha TL+\frac{1}{\mu_0}[2\cdot25W-R_3-0\cdot5R_4]$	$0\cdot25$	
HG	$\alpha TL+\frac{1}{\mu_0}[1\cdot5W-0\cdot5\sqrt{2}.R_1-R_3-R_4]$	$0\cdot5$	
GF	$\alpha TL+\frac{1}{\mu_0}[1\cdot5W-0\cdot5\sqrt{2}.R_2-R_3-R_4]$	$0\cdot5$	$0\cdot5$
FE	$\alpha TL+\frac{1}{\mu_0}[0\cdot75W-R_3-0\cdot5R_4]$	$0\cdot75$	$0\cdot5$
BH	$\alpha TL+\frac{1}{\mu_0}[1\cdot25W-0\cdot5\sqrt{2}.R_1-0\cdot5R_4]$	$0\cdot25$	
CG	$\alpha TL+\frac{1}{\mu_0}[-0\cdot5\sqrt{2}.R_1-0\cdot5\sqrt{2}.R_2-R_4]$	0	
DF	$\alpha TL+\frac{1}{\mu_0}[0\cdot75W-0\cdot5\sqrt{2}.R_2-0\cdot5R_4]$	$0\cdot75$	1
BG	$\sqrt{2}.\alpha TL+\frac{1}{\mu_0}R_1$	0	
CH	$\sqrt{2}.\alpha TL+\frac{1}{\mu_0}[0\cdot75\sqrt{2}.W+R_1+0\cdot5\sqrt{2}.R_4]$	$-0\cdot25\sqrt{2}$	
DG	$\sqrt{2}.\alpha TL+\frac{1}{\mu_0}R_2$	0	$-0\cdot5\sqrt{2}$
CF	$\sqrt{2}.\alpha TL+\frac{1}{\mu_0}[-0\cdot75\sqrt{2}.W+R_2+0\cdot5\sqrt{2}.R_4]$	$0\cdot25\sqrt{2}$	
G	$-\frac{2}{\mu_0}R_4$	0	$-0\cdot5$

It is often possible to make use of this fact to shorten the labour of computation. In this particular example, the position of F is defined by the statically determinate arrangement of bars shown in Fig. 5.2(c). The bar forces P'' due to a unit vertical load at F are given in Table 5.2, and using equation (5.4) it is found that

$$v_F = \frac{1}{\mu_0}[2{\cdot}625W - 1{\cdot}25\sqrt{2}\,.\,R_2 - R_3 - 0{\cdot}75R_4]$$

Substituting for the redundancies it is again found that

$$v_F = 0{\cdot}6316W/\mu_0 - 1{\cdot}0233\alpha TL$$

5.4. FIRST THEOREM OF COMPLEMENTARY ENERGY

It has already been pointed out that the First Theorem of Complementary Energy, equation (5.2), can be used to calculate deflections. This theorem leads to calculations which are identical with those which are made when the virtual work method is used; a general proof of this equivalence for trusses will now be given.

Consider a truss with N redundancies, $R_1, R_2, \ldots, R_n$. The actual force P in a typical bar can be expressed by the Principle of Superposition of forces as:

$$P = P_W + r_1R_1 + r_2R_2 + \ldots + r_nR_n + WP' \qquad (5.6)$$

In this equation P_W is the force due to the actual applied loads with all the redundancies and the hypothetical load W zero. The coefficients $r_1, r_2, \ldots, r_n$ have the same significance as in Chapter 4, W is a hypothetical load applied at a joint, and P' is the force in the typical bar due to unit value of W with all the applied loads and the redundancies zero.

The complementary energy c for a bar was defined by equation (3.18) as

$$c = P\lambda + \int_0^P e\,dP$$

Consider now the partial derivative $\partial c/\partial W$. This partial differentiation represents the imaginary process of varying the hypothetical load W while keeping the actual applied loads and the redundancies constant. It will of course be realised that if W were an actual load, variation of W would cause changes in the redundancies. Bearing this in mind, it follows from equation (5.6) that

$$\frac{\partial c}{\partial W} = \frac{\mathrm{d}c}{\mathrm{d}P}\frac{\partial P}{\partial W} = P'\frac{\mathrm{d}c}{\mathrm{d}P}$$

From the definition of c it follows that (equation (3.20))

$$\frac{\mathrm{d}c}{\mathrm{d}P} = \lambda + e = \delta l,$$

where δl is the total elongation of the bar. Hence

$$\frac{\partial c}{\partial W} = P'\delta l,$$

and summing over all the bars,

$$\frac{\partial C}{\partial W} = \sum_{\text{bars}} P'\delta l$$

The First Theorem of Complementary Energy, equation (5.2), which states that the deflection y corresponding to W is $\partial C/\partial W$, thus reduces to

$$y = \sum_{\text{bars}} P'\delta l,$$

and this is precisely the result obtained immediately by the virtual work transformation.

5.5. CASTIGLIANO'S THEOREM (PART II)

This theorem (Castigliano, 1879) is the special form which the First Theorem of Complementary Energy takes when the structure is linear elastic, and in addition the extension of each member from the original datum configuration is due entirely to the load in the member. Under these circumstances, λ is zero for each member, so that

$$c = \int_0^P e\mathrm{d}P,$$

and moreover, if the member characteristic is $P = \mu e$,

$$c = \int_0^P \frac{P}{\mu}\,\mathrm{d}P = P^2/2\mu$$

The strain energy u is given by

$$u = \int_0^e P\,\mathrm{d}e = \int_0^e \mu e\,\mathrm{d}e = \tfrac{1}{2}\mu e^2 = P^2/2\mu$$

Thus for each member, the strain and complementary energies are equal, so that for the whole truss, $U = C$. It follows that the deflection Δ corresponding to a load W will be given by:

$$\Delta = \frac{\partial U}{\partial W} \tag{5.7}$$

In this statement of the theorem, the symbol Δ rather than y is used to denote the deflection corresponding to W; this serves as a reminder that equation (5.7) can only be used to calculate deflections caused by applied loads acting on linear elastic structures. On the other hand, the First Theorem of Complementary Energy, namely

$$y = \frac{\partial C}{\partial W}$$

can be used to calculate deflections irrespective of the cause of the distortions of the members.

Both of these energy theorems can be applied to other types of structure than the pin-jointed truss, but proofs will not be given here.

5.6. FURTHER EXAMPLES

Simply supported beam with uniformly distributed load

Figure 5.3(a) shows a uniform simply supported beam of span L and relevant flexural rigidity EI carrying a uniformly distributed load of intensity w. It is required to determine the deflected form of the beam.

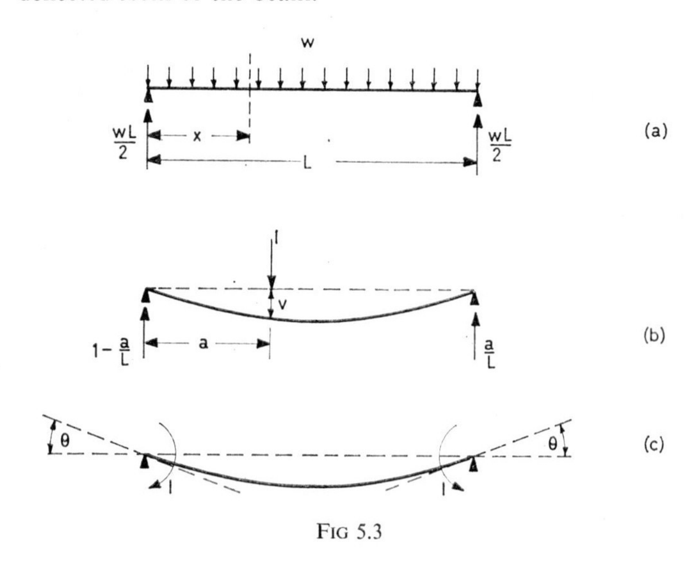

FIG 5.3

This problem could of course be solved quite easily by the direct geometrical approach of integrating the usual flexural equation

$$EI\,\frac{d^2v}{dx^2} = \text{hogging moment},$$

v being the vertical deflection at a distance x from a support, as shown. However, it will now be solved by the virtual work method, and also equivalently by Castigliano's Theorem (Part II).

The hogging bending moment M at a distance x from the left-hand support is given by

$$M = -\tfrac{1}{2}wx(L-x), \tag{5.8}$$

so that the hogging curvature κ at this section is

$$\kappa = \frac{M}{EI} = -\frac{wx}{2EI}(L-x) \tag{5.9}$$

Equation (5.9) gives the actual distribution of curvature throughout the beam; this curvature distribution, which must satisfy the requirements of compatibility, will be used in the virtual work equation, together with the actual deflection v at a distance a from the left-hand support.

The hypothetical force system consists of a unit vertical load at the section at distance a from the left-hand support, where the value of the deflection v is sought, as in Fig. 5.3(b). The hypothetical bending moments M^* which are in equilibrium with this load are as follows:

$$\left.\begin{aligned} M^* &= -\left(1-\frac{a}{L}\right)x \qquad \text{for } x \leqslant a \\ M^* &= -\frac{a}{L}(L-x) \qquad \text{for } x \geqslant a \end{aligned}\right\} \tag{5.10}$$

The virtual work equation was given in Section 3.2 (equation (3.6)) as

$$\sum_{\text{joints}} (H^*h^{**}+V^*v^{**}+C^*\phi^{**}) + \oint (w_n^*y_n^{**}+w_t^*y_t^{**})\,\mathrm{d}s$$
$$= \sum_{\text{hinges}} M^*\theta^{**} + \oint (M^*\kappa^{**}+P^*\varepsilon^{**})\,\mathrm{d}s$$

The only external load in the hypothetical load system is the unit vertical load, where the corresponding actual deflection is v, together with the end reactions $(1-a/L)$ and a/L shown in Fig. 5.3(b), for which the corresponding deflections are zero. The left-hand side of this equation therefore reduces to $1 \times v$.

At each end of the beam, $M^* = 0$, and P^* is zero at every section. Thus the only non-zero term on the right-hand side of the equation is that involving $M^*\kappa^{**}$. Substituting the values of M^* from equations (5.10), it follows that

$$1.v = \int_0^a -\left(1-\frac{a}{L}\right)x\kappa^{**}\,dx + \int_a^L -\frac{a}{L}(L-x)\kappa^{**}\,dx$$

Now $\kappa^{**} = \kappa$, and so making use of equation (5.9)

$$v = \int_0^a \left(1-\frac{a}{L}\right)x\frac{wx}{2EI}(L-x)\,dx + \int_a^L \frac{a}{L}(L-x)\frac{wx}{2EI}(L-x)\,dx \qquad (5.11)$$

$$= wa(L-a)(L^3+L^2a-La^2)/24EIL \qquad (5.12)$$

Solution by Castigliano's Theorem (Part II)

Since the beam is linear elastic and the deflections are due solely to applied load, Castigliano's Theorem (Part II) may also be used to determine the deflected form.

The strain energy δu in an element of beam of length δx was defined in equation (3.15) as

$$\frac{du}{dx} = \int_0^\kappa M\,d\kappa = \int_0^\kappa EI\kappa\,d\kappa = \tfrac{1}{2}EI\kappa^2,$$

so that for the whole beam, the strain energy U is given by

$$U = \int_0^L \tfrac{1}{2}EI\kappa^2\,dx \qquad (5.13)$$

With the distributed loading of intensity w together with a (hypothetical) concentrated load W at a distance a from the left-hand support, the hogging curvature κ at any section is found from equations (5.9) and (5.10) to be

$$\kappa = -\frac{wx}{2EI}(L-x) - \frac{Wx(L-a)}{EIL} \qquad \text{for } x \leqslant a$$

$$\kappa = -\frac{wx}{2EI}(L-x) - \frac{Wa(L-x)}{EIL} \qquad \text{for } x \geqslant a$$

It follows from equation (5.13) that

$$U = \int_0^a \tfrac{1}{2}EI\left[-\frac{wx}{2EI}(L-x)-\frac{Wx(L-a)}{EIL}\right]^2 dx$$

$$+\int_a^L \tfrac{1}{2}EI\left[-\frac{wx}{2EI}(L-x)-\frac{Wa(L-x)}{EIL}\right]^2 dx \quad (5.14)$$

Castigliano's Theorem (Part II), equation (5.7), now states that the deflection Δ corresponding to W is given by

$$\Delta = \frac{\partial U}{\partial W}$$

Differentiating equation (5.14)

$$\Delta = \frac{\partial U}{\partial W} =$$

$$\int_0^a EI\left[-\frac{wx}{2EI}(L-x)-\frac{Wx(L-a)}{EIL}\right]\left[-\frac{x(L-a)}{EIL}\right]dx$$

$$+\int_a^L EI\left[-\frac{wx}{2EI}(L-x)-\frac{Wa(L-x)}{EIL}\right]\left[-\frac{a(L-x)}{EIL}\right]dx$$

This expression gives the value of Δ for any value of W; in the actual problem W is zero, so that the required deflection v is given by

$$v = \int_0^a \frac{wx}{2EI}(L-x)x\left(1-\frac{a}{L}\right)dx+\int_a^L \frac{wx}{2EI}(L-x)\frac{a}{L}(L-x)\,dx$$

It will be seen that this is identical with equation (5.11), which was obtained by virtual work.

Determination of end slope

Another illustration of the application of virtual work is afforded by the problem of determining the slope θ at each end of the beam. The appropriate hypothetical force system consists of equal and opposite couples of unit magnitude applied at the ends of the beam, as shown in Fig. 5.3(c). The hypothetical bending moment distribution is then simply a uniform sagging moment of value unity at any section, and the virtual work equation is

$$1.\theta + 1.\theta = \int_0^L (-1)\kappa \, dx$$

$$= \int_0^L \frac{wx}{2EI}(L-x)\, dx,$$

using equation (5.9). Performing the integration, it follows that

$$\theta = \frac{wL^3}{24EI},$$

a result which could of course have been obtained by direct differentiation of equation (5.12).

Deflections at plastic collapse

The virtual work approach can also be used to advantage for estimating the deflections of beams and plane frames at plastic collapse (Heyman, 1961). Figure 5.4(a) shows a uniform beam AC of length L which is rigidly built in at C and rests on a simple support at A. The beam is subjected to a concentrated load W at its mid-point B. The beam has flexural rigidity EI and fully plastic moment M_p, and it is required to find the deflection at B when W has just attained the value at which plastic collapse occurs, but before any motion of the collapse mechanism has taken place. It is assumed that the beam is wholly elastic between the plastic hinges.

For this beam it is evident that plastic collapse will occur when plastic hinges have formed at B and C. The collapse mechanism is depicted in Fig. 5.4(b), and the corresponding bending moment distribution is given in Fig. 5.4(c). The value

W_c of W at which collapse occurs may be found by considering the kinematics of the collapse mechanism, as discussed in

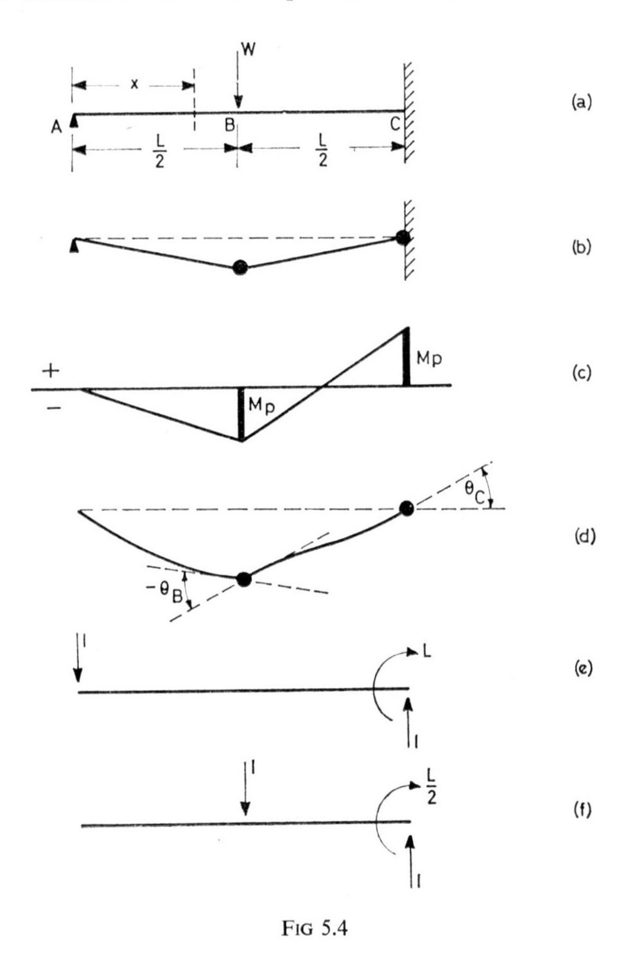

FIG 5.4

Chapter 8, or by noting from Fig. 5.4(c) that the shear forces in AB and BC, which are equal to the corresponding rates of change of bending moment, are $2M_p/L$ and $4M_p/L$, respectively.

Thus for vertical equilibrium

$$W_c = \frac{2M_p}{L} + \frac{4M_p}{L} = \frac{6M_p}{L} \tag{5.15}$$

As the load W is increased steadily from zero, the beam at first remains wholly elastic. At some value of W the first plastic hinge will form, but it is not known *a priori* whether this will occur at B or C. With further increase of the load, this hinge will undergo rotation, while the bending moment at the other possible hinge position increases. Eventually, the fully plastic moment is attained at this second hinge position when $W = W_c$; the deflection at B which is required is the deflection at this value of the load but before the second hinge has undergone any rotation.

The analysis involves using the actual distribution of curvature in the beam in conjunction with certain hypothetical distributions of bending moment. From Fig. 5.4(c) it is seen that at collapse the hogging bending moment M at a distance x from A is given by

$$M = -2xM_p/L \qquad \text{for} \quad 0 \leqslant x \leqslant L/2$$

$$M = (4x-3L)M_p/L \qquad \text{for } L/2 \leqslant x \leqslant L$$

The actual hogging curvature κ is then given by M/EI, so that

$$\left.\begin{aligned} \kappa &= -2xM_p/EIL && \text{for} \quad 0 \leqslant x \leqslant L/2 \\ \kappa &= (4x-3L)M_p/EIL && \text{for } L/2 \leqslant x \leqslant L \end{aligned}\right\} \tag{5.16}$$

These expressions define the curved shape of the beam in the segments AB and BC; in addition, at the point of collapse there will be a hinge rotation at the position of the first hinge to form. Since this position is as yet unknown, hinge rotations $-\theta_B$ and θ_C are shown in the deflected form of Fig. 5.4(d). The signs of these hinges are chosen to correspond to the bending moments acting at each of these two sections; at B the bending moment is sagging so that a sagging hinge must form at that section; and similarly the hinge at C must be hogging.

The actual deformations defined above are first used in the virtual work equation in conjunction with the hypothetical force system of Fig. 5.4(e), consisting of equal and opposite unit vertical loads at A and C together with a counter-clockwise moment L at C. These external loads and the moment are in equilibrium with a hogging bending moment of magnitude x at a distance x from A. This hypothetical force system is chosen in such a way that the external loads and the moment do no virtual work on the actual deflections of the beam, none of which are known at this stage. What is sought at this stage is information concerning the hinge rotations θ_B and θ_C.

Since the virtual work done is zero, the virtual work equation reduces to

$$\sum_{\text{hinges}} M^*\theta^{**} + \oint M^*\kappa^{**}\,ds = 0$$

Remembering that $M^* = x$, and that $\kappa^{**} = \kappa$, the actual curvature at any section, this equation becomes

$$\tfrac{1}{2}L(-\theta_B) + L\theta_C + \int_0^L x\kappa\,dx = 0$$

Substituting the values of κ from equations (5.16), this gives

$$-\tfrac{1}{2}L\theta_B + L\theta_C + \int_0^{L/2} x\left[-\frac{2xM_p}{EIL}\right]dx + \int_{L/2}^{L} x\left[\frac{(4x-3L)M_p}{EIL}\right]dx = 0,$$

from which it follows that

$$\theta_C = \tfrac{1}{2}\theta_B + \frac{M_pL}{24EI} \tag{5.17}$$

This is the condition of compatibility which must be fulfilled in the collapse condition.

The signs of the hinge rotations were chosen to be consistent with the fully plastic moments at the hinges, so that θ_C and θ_B must both be positive. If the last hinge were supposed to form at C, so that at the point of collapse θ_C was zero, it is seen from equation (5.17) that θ_B would then be negative, which is impossible. It is therefore concluded that the last hinge forms at B, so that from equation (5.17) the situation at the point of collapse is

$$\theta_B = 0, \qquad \theta_C = \frac{M_p L}{24EI} \tag{5.18}$$

The deflection at B is now obtained in the usual way by using the actual deformations in the virtual work equation together with a hypothetical force system which satisfies the requirements of equilibrium with a unit vertical load at B. One such system is illustrated in Fig. 5.4(f); the bending moment distribution is:

$$\begin{aligned} M^* &= 0 && \text{for} \quad 0 \leqslant x \leqslant L/2 \\ M^* &= x - L/2 && \text{for } L/2 \leqslant x \leqslant L \end{aligned}$$

Thus if v_B is the vertical deflection at B, the virtual work equation is

$$\begin{aligned} 1.v_B &= \frac{L}{2}\,\theta_C + \int_{L/2}^{L} (x - L/2)\,\kappa\,\mathrm{d}x \\ &= \frac{M_p L^2}{48EI} + \int_{L/2}^{L} (x - L/2)\,\frac{(4x - 3L)M_p}{EIL}\,\mathrm{d}x, \end{aligned}$$

using equations (5.16) and (5.18). Evaluating the integral, it is found that

$$v_B = \frac{M_p L^2}{16EI},$$

so that from equation (5.15)

$$v_B = \frac{W_c L^3}{96EI}$$

Further examples of the application of this technique to more complex beam and frame problems have been given by Heyman (1964).

6

Indeterminate structures by the equilibrium method

6.1. INTRODUCTION

In analysing indeterminate structures it is first necessary to choose the most appropriate variables, as pointed out in Chapter 1. These may be force variables, or the redundancies, which are found from the equations of geometrical compatibility, as described in Chapter 4. Alternatively deformation variables may be used, and these must be found by establishing appropriate equations of equilibrium. The present chapter is concerned with this latter approach, which is often termed the equilibrium method.

The equilibrium method is generally more convenient than the compatibility approach for those problems in which the number of deformation variables is less than the number of force variables or redundancies, although different factors may govern the choice of method if computer programmes are to be used (Livesley, 1964). For most pin-jointed trusses there are fewer force variables, so that the compatibility method is usually simpler. However, it is easier to explain the procedure with reference to truss problems, and so the first examples in this chapter relate to simple trusses. The chapter concludes with a discussion of the equilibrium method in relation to plane frames.

Each problem is analysed by an indirect method, so that the

equilibrium equations are not established by applications of the principles of statics. The Principle of Virtual Work is used to transform the actual equilibrium problem into a hypothetical geometrical problem of determining systems of deformations which satisfy the requirements of compatibility with certain hypothetical joint displacements or rotations.

Some of the energy theorems are also shown to achieve identical transformations. These theorems, which are also discussed in relation to those problems to which they are applicable, are

Theorem of Minimum Potential Energy,
Castigliano's Theorem (Part I), and
First Theorem of Minimum Strain Energy.

6.2. TRUSS WITH TWO DEFORMATION VARIABLES

Consider the truss whose dimensions and loading are as illustrated in Fig. 6.1(a). It is required to determine the stresses in the bars, and the horizontal and vertical displacements Δ_1 and Δ_2 of F, due to the applied loads, the truss being free from stress when unloaded.

The relationship between the force P in a typical bar and its extension e due to this force is supposed to be

$$P = f(e) \tag{6.1}$$

where $f(e)$ is a single-valued function of e. Particular solutions for two cases will be given, the first solution being for a linear relationship $P = \mu e$ and the second for a non-linear relationship.

The truss has three redundancies, so that if the compatibility method was used it would be necessary to establish three compatibility equations. However, there are only two deformation variables, namely the horizontal and vertical components of displacement of F. Hence it would appear that the equilibrium method, involving the solution of two equilibrium equations for the determination of Δ_1 and Δ_2, can be used with advantage.

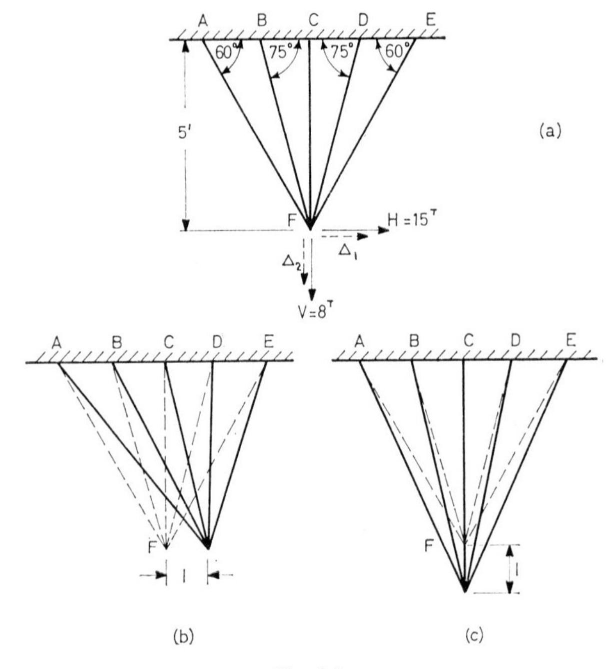

FIG 6.1

Compatibility

The first step in the analysis is to express the extension e of each bar in terms of the two unknown deflections Δ_1 and Δ_2. This is carried out most conveniently by determining the extensions of the bars in the two displacement systems illustrated in Figs. 6.1(b) and (c). These are defined as follows:

(i) $\Delta_1 = 1, \Delta_2 = 0; \quad e = a_1$

(ii) $\Delta_1 = 0, \Delta_2 = 1; \quad e = a_2$

By the Principle of Superposition of displacements the actual extension of any bar is then given by

$$e = a_1\Delta_1 + a_2\Delta_2 \tag{6.2}$$

Member characteristics

The force P in each bar is then found from the member characteristic, equation (6.1), to be

$$P = f(e) = f(a_1\Delta_1 + a_2\Delta_2) \tag{6.3}$$

Equilibrium

There now remains the problem of determining the correct values of Δ_1 and Δ_2 which are such that the conditions of equilibrium are satisfied. In fact, there are two equations of equilibrium which could be obtained directly by resolving horizontally and vertically for equilibrium of joint F. However, these equations will be derived by using the Principle of Virtual Work to transform this actual problem of statical equilibrium into a hypothetical geometrical problem.

The transformation is achieved by using the actual bar forces P in the Principle of Virtual Work, since these are known to satisfy the requirements of equilibrium with the applied loads. Two hypothetical displacement systems are used in turn, these being the bar extensions a_1 and a_2 which are compatible with unit horizontal and vertical displacements of F, respectively.

The virtual work equation for trusses was given in Chapter 3, equation (3.1a), as

$$\sum_{\text{joints}} (H^*h^{**} + V^*v^{**}) = \sum_{\text{bars}} P^*(\delta l)^{**}$$

In both the displacement systems, the deflections of joints A, B, C, D and E are all zero, so that the only terms involved in the summation on the left-hand side of this equation are for joint F. In the first displacement system, the horizontal deflection of F is unity while the vertical deflection is zero, so that this summation reduces to $1 \times H^*$, or simply H, where $H = 15$ ton is the actual horizontal load at F.

Turning now to the right-hand side of the virtual work equation, the extension of a typical bar which is compatible with unit horizontal deflection of F has already been defined as a_1. The actual force P in any bar is given by equation (6.3) as $f(a_1\Delta_1 + a_2\Delta_2)$. Thus a typical term in the summation is $a_1 f(a_1\Delta_1 + a_2\Delta_2)$, so that the virtual work equation becomes

$$H = \sum_{\text{bars}} a_1 f(a_1\Delta_1 + a_2\Delta_2) = \sum_{\text{bars}} a_1 P \tag{6.4}$$

Using the second displacement system, it can be shown similarly that

$$V = \sum_{\text{bars}} a_2 f(a_1\Delta_1 + a_2\Delta_2) = \sum_{\text{bars}} a_2 P \tag{6.5}$$

Equations (6.4) and (6.5) are respectively the equations of horizontal and vertical equilibrium for the joint F.

Solution for elastic stress/strain relation

For a numerical solution it will first be supposed that each bar in the truss has a cross-sectional area A_0 of 1 sq. in., and that the relationship between stress σ and strain ε is

$$\sigma = E\varepsilon, \tag{6.6}$$

where $E = 4500$ ton/in^2. The relationship between bar force P (ton) and extension e (in.) is then

$$P = \mu e = \mu(a_1\Delta_1 + a_2\Delta_2), \tag{6.7}$$

making use of equation (6.2). In this equation, μ is the stiffness of the bar, and is given by

$$\mu = \frac{EA_0}{l_0} = \frac{4500}{l_0}, \tag{6.8}$$

where l_0 is the original length of the bar.

Substituting equation (6.7) in the equilibrium equations (6.4) and (6.5), it is found that

$$\left.\begin{aligned} H &= \Delta_1 \sum \mu a_1^2 + \Delta_2 \sum \mu a_1 a_2 \\ V &= \Delta_1 \sum \mu a_2 a_1 + \Delta_2 \sum \mu a_2^2 \end{aligned}\right\} \tag{6.9}$$

Values of l_0, a_1, a_2, and the three quantities involved in the summations of equations (6.9) are given in Table 6.1.

TABLE 6.1

Member	l_0(*in.*)	a_1	a_2	μa_1^2	$\mu a_1 a_2$	μa_2^2	σ(*ton/in*^{2})
AF	69·28	0·5000	0·8660	16·24	28·13	48·71	13·01
BF	62·12	0·2588	0·9659	4·85	18·11	67·59	8·49
CF	60	0	1	0	0	75	1·95
DF	62·12	−0·2588	0·9659	4·85	−18·11	67·59	−4·85
EF	69·28	−0·5000	0·8660	16·24	−28·13	48·71	−10·09
			$\sum_{\text{bars}}$	42·18	0	307·60	

Substituting the values of the summations in equations (6.9), it is found that

$$H = 15 = 42{\cdot}18\Delta_1$$

$$V = 8 = 307{\cdot}60\Delta_2,$$

from which

$$\Delta_1 = 0{\cdot}3556 \text{ in.}$$

$$\Delta_2 = 0{\cdot}0260 \text{ in.}$$

Once these values of Δ_1 and Δ_2 are known, the stresses in the bars are readily found from equations (6.6) and (6.7), and the values of σ are given in the last column of Table 6.1.

It will be seen that the coefficients of Δ_1 and Δ_2 in equations (6.9) form a symmetrical matrix, which may be referred to as the stiffness matrix, since each coefficient has the dimensions of a stiffness. The form of this matrix should be compared with that of the flexibility matrix occurring in the compatibility equations obtained in Chapter 4, equations (4.9). According to Argyris and Kelsey (1960), the general procedure for setting up the equilibrium equations in terms of stiffness matrices was first established by Ostenfeld (1926).

Solution for non-linear stress/strain relation

The problem of Fig. 6.1 will now be solved for the case in which the relationship between stress and strain for each bar is the non-linear relation

$$\sigma = 4500\varepsilon - 55{\cdot}10^6\varepsilon^3 \tag{6.10}$$

This stress/strain relation is shown in Fig. 6.2; it will be seen that it corresponds to a 0·1 per cent. proof stress of 15 ton/in².

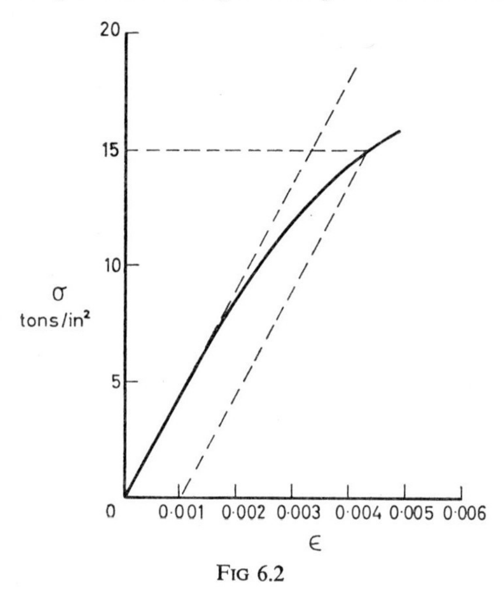

FIG 6.2

The two equilibrium equations (6.4) and (6.5) are still valid, and the coefficients a_1 and a_2 in Table 6.1 are unaffected; the

E

only change is in the form of $f(e)$. For any bar, $\varepsilon = e/l_0$, so that the relationship between P and e is given by

$$P = \sigma = 4500\varepsilon - 55{\cdot}10^6\varepsilon^3$$

$$= 4500\left(\frac{e}{l_0}\right) - 55{\cdot}10^6\left(\frac{e}{l_0}\right)^3$$

$$= \frac{4500}{l_0}(a_1\Delta_1 + a_2\Delta_2) - \frac{55{\cdot}10^6}{l_0^3}(a_1\Delta_1 + a_2\Delta_2)^3 \tag{6.11}$$

Values of P obtained from equation (6.11) and the data of Table 6.1 are set out in Table 6.2.

TABLE 6.2

Member	*P(ton)*	*σ(ton/in²)*
AF	$(32{\cdot}48\Delta_1 + 56{\cdot}25\Delta_2) - 165{\cdot}4(0{\cdot}5000\Delta_1 + 0{\cdot}8660\Delta_2)^3$	12·57
BF	$(18{\cdot}75\Delta_1 + 69{\cdot}97\Delta_2) - 229{\cdot}4(0{\cdot}2588\Delta_1 + 0{\cdot}9659\Delta_2)^3$	8·92
CF	$75\Delta_2 - 254{\cdot}6\Delta_2^3$	2·25
DF	$(-18{\cdot}75\Delta_1 + 69{\cdot}97\Delta_2) - 229{\cdot}4(-0{\cdot}2588\Delta_1 + 0{\cdot}9659\Delta_2)^3$	−5·12
EF	$(-32{\cdot}48\Delta_1 + 56{\cdot}25\Delta_2) - 165{\cdot}4(-0{\cdot}5000\Delta_1 + 0{\cdot}8660\Delta_2)^3$	−10·17

The resulting equations of equilibrium are

$$H = 15 = 42{\cdot}18\Delta_1 - 82{\cdot}7[(0{\cdot}5000\Delta_1 + 0{\cdot}8660\Delta_2)^3 - (-0{\cdot}5000\Delta_1 + 0{\cdot}8660\Delta_2)^3] - 59{\cdot}4[(0{\cdot}2588\Delta_1 + 0{\cdot}9659\Delta_2)^3 - (-0{\cdot}2588\Delta_1 + 0{\cdot}9659\Delta_2)^3]$$

$$V = 8 = 307{\cdot}6\Delta_2 - 254{\cdot}6\Delta_2^3 - 143{\cdot}2[(0{\cdot}5000\Delta_1 + 0{\cdot}8660\Delta_2)^3 + (-0{\cdot}5000\Delta_1 + 0{\cdot}8660\Delta_2)^3] - 221{\cdot}6[(0{\cdot}2588\Delta_1 + 0{\cdot}9659\Delta_2)^3 + (-0{\cdot}2588\Delta_1 + 0{\cdot}9659\Delta_2)^3]$$

The solution of these equations is

$$\Delta_1 = 0{\cdot}3899 \text{ in.}$$

$$\Delta_2 = 0{\cdot}0301 \text{ in.},$$

and the corresponding values of the stresses are given in the last column of the table.

Solution by Theorem of Minimum Potential Energy

The equations of equilibrium for this problem could also have been derived by applying the Theorem of Minimum Potential Energy. It will be recalled from Chapter 3 that the total potential ϕ is defined as

$$\phi = U+V$$

In this expression U represents the strain energy expressed in terms of deformation variables, and V is the potential energy of the external loads, being given by

$$V = -\sum_{\text{joints}} (Hh+Vv)$$

The Theorem of Minimum Potential Energy states that amongst all possible sets of deformations, those which ensure that all the conditions of equilibrium are fulfilled are those for which the total potential ϕ is minimised. In the particular problem under consideration, there are two deformation variables Δ_1 and Δ_2, so that according to this theorem the equilibrium equations are

$$\frac{\partial\phi}{\partial\Delta_1} = 0, \qquad \frac{\partial\phi}{\partial\Delta_2} = 0$$

The strain energy u for a bar was defined in Chapter 3, equation (3.12), as

$$u = \int_0^e P\,\text{d}e$$

For the elastic stress/strain law $\sigma = E\varepsilon$, the (P,e) relation for a typical bar was given in equation (6.7) as

$$P = \mu e,$$

so that

$$u = \int_0^e \mu e \, de$$

$$= \tfrac{1}{2}\mu e^2$$

$$= \tfrac{1}{2}\mu(a_1\Delta_1 + a_2\Delta_2)^2,$$

making use of equation (6.2). Thus for the whole truss

$$U = \sum_{\text{bars}} \tfrac{1}{2}\mu(a_1\Delta_1 + a_2\Delta_2)^2$$

The potential energy V of the applied loads is seen from Fig. 6.1(a) to be

$$V = -H\Delta_1 - V\Delta_2$$

It follows that

$$\phi = U + V = \sum_{\text{bars}} \tfrac{1}{2}\mu(a_1\Delta_1 + a_2\Delta_2)^2 - H\Delta_1 - V\Delta_2$$

The conditions $\partial\phi/\partial\Delta_1 = 0$ and $\partial\phi/\partial\Delta_2 = 0$ are then seen at once to be

$$\sum_{\text{bars}} a_1\mu(a_1\Delta_1 + a_2\Delta_2) - H = 0$$

$$\sum_{\text{bars}} a_2\mu(a_1\Delta_1 + a_2\Delta_2) - V = 0$$

or making use of equation (6.7),

$$H = \sum_{\text{bars}} a_1 P$$

$$V = \sum_{\text{bars}} a_2 P$$

These are precisely equations (6.4) and (6.5) which were derived by virtual work.

There is little difficulty in showing that the conditions $\partial\phi/\partial\Delta_1 = 0$ and $\partial\phi/\partial\Delta_2 = 0$ also reduce to equations (6.4) and (6.5) if the stress/strain relation is the non-linear law of equation (6.10).

6.3. LACK OF FIT

The method of analysis described in Section 6.2 is immediately applicable to cases in which there is initial lack of fit.

Consider for example the truss of Fig. 6.1(a), and suppose that the bars AF, BF, CF and EF are first assembled with DF omitted, and that these four bars are then free from stress. The bar DF is too long to fit in freely by an amount λ_0 equal to 0·1 in. It is required to find the stresses which would be set up if DF was forced into place, there being no external load at F after the force fit had occurred.

Compatibility

Suppose that the final horizontal and vertical deflections of F are y_1 and y_2, these being measured from the datum position of F before the bar DF was inserted. Then the total extension δl of any bar, measured from this datum, is given by

$$\delta l = a_1 y_1 + a_2 y_2,$$

where a_1 and a_2 have the same significance as in equation (6.2).

δl is the sum of the extension e due to axial load P and the extension λ due to all other causes, so that

$$\begin{aligned} e &= \delta l - \lambda \\ &= a_1 y_1 + a_2 y_2 - \lambda \end{aligned} \qquad (6.12)$$

In this particular case, λ is zero for all the bars except DF, for which it is equal to the lack of fit λ_0.

Member characteristics

If the relationship between axial force P and the elongation e due to this force is $P = f(e)$, then it follows from equation (6.12) that

$$P = f(e) = f(a_1 y_1 + a_2 y_2 - \lambda) \qquad (6.13)$$

Equilibrium

The forces P given by equation (6.13) are the actual bar forces and so satisfy the requirements of equilibrium with zero external load at F. These forces will be used in the virtual work equation.

As before, two hypothetical displacement systems are used in the virtual work equation, these consisting of the bar extensions a_1 which are compatible with unit horizontal displacement of F, and the bar extensions a_2 which are compatible with unit vertical displacement of F. Since there is no applied load at F, and no displacements at A, B, C, D or E, the virtual work done in each case is zero, and it follows that

$$\sum_{\text{bars}} a_1 f(a_1 y_1 + a_2 y_2 - \lambda) = \sum_{\text{bars}} a_1 P = 0 \qquad (6.14)$$

$$\sum_{\text{bars}} a_2 f(a_1 y_1 + a_2 y_2 - \lambda) = \sum_{\text{bars}} a_2 P = 0 \qquad (6.15)$$

These two equations of equilibrium hold true for any relationship between P and e. As an illustration, suppose that the stress/strain relation for each bar is the elastic relation of equation (6.6), namely

$$\sigma = 4500\varepsilon,$$

where σ is in ton/in². The cross-sectional area A_0 of each bar will be taken as 1 sq. in., as before, so that the (P,e) relation is

$$P = \sigma = \mu e = \mu(a_1 y_1 + a_2 y_2 - \lambda) \qquad (6.16)$$

Substituting equation (6.16) in equations (6.14) and (6.15), it is found that

$$\left.\begin{aligned} y_1 \sum \mu a_1^2 + y_2 \sum \mu a_1 a_2 &= \sum \mu a_1 \lambda \\ y_1 \sum \mu a_2 a_1 + y_2 \sum \mu a_2^2 &= \sum \mu a_2 \lambda \end{aligned}\right\} \qquad (6.17)$$

The stiffness coefficients appearing on the left-hand sides of these equations have already been evaluated in Table 6.1.

Since λ is zero for every member except DF, the right-hand sides of these equations become

$$\left[\frac{4500a_1\lambda}{l_0}\right]_{DF} = -\frac{4500\times 0{\cdot}2588\times 0{\cdot}1}{62{\cdot}12} = -1{\cdot}875$$

and

$$\left[\frac{4500a_2\lambda}{l_0}\right]_{DF} = \frac{4500\times 0{\cdot}9659\times 0{\cdot}1}{62{\cdot}12} = 6{\cdot}997,$$

using the data for DF given in Table 6.1. Referring again to this table for the values of the stiffness coefficients, equations (6.17) are seen to be

$$42{\cdot}18y_1 = -1{\cdot}875$$

$$307{\cdot}60y_2 = 6{\cdot}997,$$

whence

$$y_1 = -0{\cdot}04434 \text{ in.}$$

$$y_2 = 0{\cdot}02275 \text{ in.}$$

Corresponding values of the stresses in the bars, obtained from equation (6.16) are given in Table 6.3.

TABLE 6.3

Member	AF	BF	CF	DF	EF
$\sigma(ton/in^2)$	2·72	2·42	1·71	−6·48	−0·16

Equations (6.14) and (6.15) could also have been obtained by using the Theorem of Minimum Potential Energy. The derivation will not be given here; instead, it will be shown in the next section that this theorem always leads to equilibrium equations identical with those derived by the virtual work method.

6.4. THEOREM OF MINIMUM POTENTIAL ENERGY

The implications of the Theorem of Minimum Potential Energy will now be considered in general terms for pin-jointed trusses. For any truss, let there be n deformation variables $y_1, y_2, \ldots, y_q, \ldots, y_n$, these being possible independent movements of the joints. In the truss of Fig. 6.3, for example, n would be twelve. Then the elongation δl of any bar from an original datum position can be expressed as

$$\delta l = a_1 y_1 + a_2 y_2 + \ldots + a_q y_q + \ldots + a_n y_n \qquad (6.18)$$

This result follows at once from the Principle of Superposition for displacements; a_q is the elongation of the bar when $y_q = 1$ and all the other deformation variables are zero.

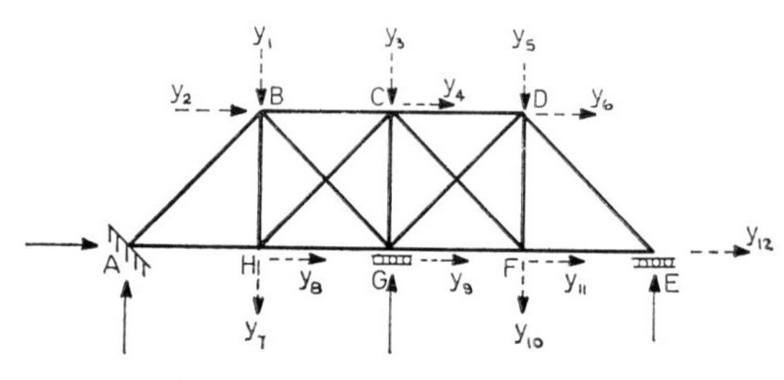

FIG 6.3

In general, $\delta l = e + \lambda$, where e is the elongation due to axial force P and λ is the elongation due to all other causes (such as lack of fit or temperature strain). Thus from equation (6.18),

$$e = \delta l - \lambda = a_1 y_1 + a_2 y_2 + \ldots + a_q y_q + \ldots + a_n y_n - \lambda \qquad (6.19)$$

If there exists a unique relationship between P and e, so that $P = f(e)$, it follows that

$$P = f(e) = f(a_1 y_1 + a_2 y_2 + \ldots + a_q y_q + \ldots + a_n y_n - \lambda) \qquad (6.20)$$

Equation (6.20) expresses the actual force P in any bar as a function of the deformation variables. The actual bar forces P must satisfy the requirements of equilibrium with the external loads on the truss, which in general will consist of n components

$W_1, W_2, \ldots, W_q, \ldots, W_n$, corresponding to the n deformation variables, together with reactions at fixed points. The term corresponding is here used in the sense that a load W_q corresponds to a deflection y_q when the directions of W_q and y_q coincide. Thus in Fig. 6.3 there may be twelve components of load corresponding to the twelve deformation variables shown, and in addition vertical reactions at E and G together with horizontal and vertical reactions at A.

The n equations of equilibrium for the determination of the deformation variables are readily established by using the virtual work equation. The actual bar forces and external loads are used in this equation, since they are known to satisfy the requirements of equilibrium. n hypothetical displacement systems are used in turn, a typical one consisting of the bar elongations a_q which are compatible with a hypothetical unit displacement $y_q = 1$, with all the other deformation variables zero. The virtual work done by the applied loads is then simply $1 \times W_q$, and equating this to the virtual work absorbed in the bars gives

$$\begin{aligned} W_q &= \sum_{\text{bars}} a_q P \\ &= \sum_{\text{bars}} a_q f(a_1 y_1 + a_2 y_2 + \ldots + a_q y_q + \ldots + a_n y_n - \lambda) \end{aligned} \tag{6.21}$$

The n equations of the type of equation (6.21) with $q = 1, 2, \ldots, n$, constitute the n equations of equilibrium.

As in previous work, a flexible support is merely treated as an additional bar of the truss with zero length.

Now consider the application of the Theorem of Minimum Potential Energy. This theorem states that the equations of equilibrium are given by

$$\frac{\partial \phi}{\partial y_q} = \frac{\partial}{\partial y_q}(U + V) = 0, \qquad q = 1, 2, \ldots, n \tag{6.22}$$

The potential energy V of the applied loads was defined in Chapter 3, equation (3.25), as

$$V = -\sum_{\text{joints}} (Hh+Vv),$$

so that in this general case

$$V = -(W_1y_1+W_2y_2+\ldots+W_qy_q+\ldots+W_ny_n) \quad (6.23)$$

It follows immediately that

$$\frac{\partial V}{\partial y_q} = -W_q,$$

so that equations (6.22) become

$$W_q = \frac{\partial U}{\partial y_q}, \qquad q = 1, 2, \ldots, n \qquad (6.24)$$

The strain energy u for a typical bar was defined in Chapter 3, equation (3.12), as

$$u = \int_0^e P\,\mathrm{d}e,$$

and it was also shown that as a consequence of this definition, equation (3.13),

$$\frac{\mathrm{d}u}{\mathrm{d}e} = P$$

From equation (6.19), it therefore follows that

$$\frac{\partial u}{\partial y_q} = \frac{\partial e}{\partial y_q}\frac{\mathrm{d}u}{\mathrm{d}e} = a_qP,$$

and summing over all the bars of the truss,

$$\frac{\partial U}{\partial y_q} = \sum_{\text{bars}} a_qP$$

Combining this result with equations (6.24), it is seen that according to the Theorem of Minimum Potential Energy the equilibrium equations are

$$W_q = \sum_{\text{bars}} a_q P, \qquad q = 1, 2, \ldots, n,$$

and these equations are identical with equations (6.21), which were obtained by virtual work.

It should be noted that the Theorem of Minimum Potential Energy does not depend on the assumption of linear elasticity. It is only necessary that there should be a unique relationship $P = f(e)$ for each bar, so that the strain energy is a function of e only, as explained in Section 3.4.

Minimum of the total potential

Equations (6.22) ensure that the total potential ϕ has a stationary value when the deformation variables have the values which ensure equilibrium, but are not sufficient to prove that this value is a minimum, as implied by the title of the theorem. In fact, if the stationary value of ϕ is a minimum, the equilibrium is stable, and if it is a maximum, the equilibrium is unstable. Thus the nature of the stationary value of ϕ is of importance in stability problems, but for the type of problem under consideration the stationary value is always a minimum (Hoff, 1956).

In the special case in which λ is zero for every bar, and in addition the load/extension relation for each bar is the elastic law $P = \mu e$, it can be shown that ϕ is in fact a minimum at the equilibrium configuration. Consider a truss subjected to n components of external load $W_1, W_2, \ldots, W_q, \ldots, W_n$, and let the actual corresponding deflections be $y_1, y_2, \ldots, y_q, \ldots, y_n$, as before. The actual extension in a typical bar is denoted by e, so that the actual force in this bar is μe.

Let y_q^{**} $(q = 1, \ldots, n)$ denote any hypothetical set of joint deflections, and let e^{**} denote the extension of the typical bar which is compatible with these deflections.

Using the actual external loads and internal bar forces in the virtual work equation, together with the actual and hypothetical deformation systems in turn, it follows that

$$\sum_{\text{joints}} Wy = \sum_{\text{bars}} \mu e^2 \tag{6.25}$$

$$\sum_{\text{joints}} Wy^{**} = \sum_{\text{bars}} \mu e e^{**}, \tag{6.26}$$

so that

$$\sum_{\text{joints}} W(y-y^{**}) = \sum_{\text{bars}} \mu e(e-e^{**}) \tag{6.27}$$

Now consider the identity

$$\sum_{\text{bars}} \tfrac{1}{2}\mu(e^{**}-e)^2 \equiv \sum_{\text{bars}} \tfrac{1}{2}\mu[(e^{**})^2-e^2+2e(e-e^{**})]$$

By virtue of equation (6.27), this becomes

$$\sum_{\text{bars}} \tfrac{1}{2}\mu(e^{**}-e)^2$$

$$= \sum_{\text{bars}} \tfrac{1}{2}\mu(e^{**})^2 - \sum_{\text{joints}} Wy^{**} - \left[\sum_{\text{bars}} \tfrac{1}{2}\mu e^2 - \sum_{\text{joints}} Wy\right]$$

Since the left-hand side of this equation cannot be less than zero, it follows that

$$\sum_{\text{bars}} \tfrac{1}{2}\mu e^2 - \sum_{\text{joints}} Wy \leqslant \sum_{\text{bars}} \tfrac{1}{2}\mu(e^{**})^2 - \sum_{\text{joints}} Wy^{**} \tag{6.28}$$

The left-hand side of this inequality is the actual total potential ϕ in the equilibrium configuration, whereas the right-hand side is the total potential ϕ associated with any hypothetical deformation system which satisfies compatibility.

Thus $$\phi \leqslant \phi^{**},$$

which establishes the result.

Upper and lower bounds

As pointed out by Prager (1961), the inequality (6.28) can be combined with the inequality (4.11) obtained in Section 4.4, which stated that

$$\sum_{\text{bars}} \frac{1}{2\mu} P^2 \leqslant \sum_{\text{bars}} \frac{1}{2\mu} (P^*)^2,$$

where P^* represents any hypothetical set of bar forces which satisfies all the requirements of equilibrium with the external loads. Inverting this inequality, and substituting $P = \mu e$, it is found that

$$\sum_{\text{bars}} \frac{1}{2\mu}(P^*)^2 \geqslant \sum_{\text{bars}} \tfrac{1}{2}\mu e^2,$$

and using equation (6.25), this becomes

$$\sum_{\text{bars}} \frac{1}{2\mu}(P^*)^2 \geqslant \sum_{\text{joints}} \tfrac{1}{2}Wy \tag{6.29}$$

Using equation (6.25) again in conjunction with the inequality (6.28),

$$-\sum_{\text{bars}} \tfrac{1}{2}Wy \leqslant \sum_{\text{bars}} \tfrac{1}{2}\mu(e^{**})^2 - \sum_{\text{joints}} Wy^{**}$$

Combining this with the inequality (6.29), the following continued inequality results

$$\sum_{\text{bars}} \frac{1}{2\mu}(P^*)^2 \geqslant \sum_{\text{joints}} \tfrac{1}{2}Wy \geqslant \sum_{\text{joints}} Wy^{**} - \sum_{\text{bars}} \tfrac{1}{2}\mu(e^{**})^2 \tag{6.30}$$

This result does not appear to have useful applications in structural analysis, but Prager (1961) has shown how the corresponding result for a solid body can be used to obtain upper and lower bounds on the stiffness of a shaft in torsion. If any structural applications were found, they would almost certainly be confined to cases in which a single external load was acting; equation (6.30) would then provide upper and lower bounds on the corresponding deflection y, and hence on the stiffness.

6.5. CASTIGLIANO'S THEOREM (PART I)

This theorem (Castigliano, 1879), is merely an alternative statement of the Theorem of Minimum Potential Energy. This latter theorem was shown in equations (6.24) to be equivalent to the result that the equilibrium equations are

$$W_q = \frac{\partial U}{\partial y_q},$$

and this is the statement of Castigliano's Theorem (Part I). Like the Theorem of Minimum Potential Energy, its application leads immediately to the set of equilibrium equations (6.21), and hence to precisely the same computations for the determination of the deformation variables as are performed if the virtual work method is used.

It should also be noted that this theorem, like the Theorem of Minimum Potential Energy, is not dependent upon the assumption of a linear relationship between bar force P and elongation e.

First Theorem of Minimum Strain Energy

This theorem is merely the special case of Castigliano's Theorem (Part I) or of the Theorem of Minimum Potential Energy, when applied to an unloaded joint. From equations (6.24) it follows that under these circumstances

$$\frac{\partial U}{\partial y_q} = 0$$

6.6. BEAMS AND FRAMES

The equilibrium method is rarely used directly for the solution of problems concerning beams and frames. Instead, a wide variety of special techniques for dealing with flexural problems has been developed. However, it is instructive to see how the equilibrium method applies to such problems, since it becomes apparent that some of these techniques, in particular the moment distribution method (Hardy Cross, 1930) are based essentially upon an equilibrium approach.

The equilibrium method will be discussed in relation to the frame whose dimensions and loading are as illustrated in Fig. 6.4(a). All the members of this frame are uniform and elastic, with flexural rigidity EI, and it is required to determine the distribution of bending moments due to the applied loads, the frame being free from stress when unloaded.

By the Principle of Superposition the effect of the applied loading of Fig. 6.4(a) is equivalent to the sum of the effects of the two loading systems shown in Figs. 6.4(b) and (c). In Fig. 6.4(b) the hogging end moments $WL/4$ acting on the beam

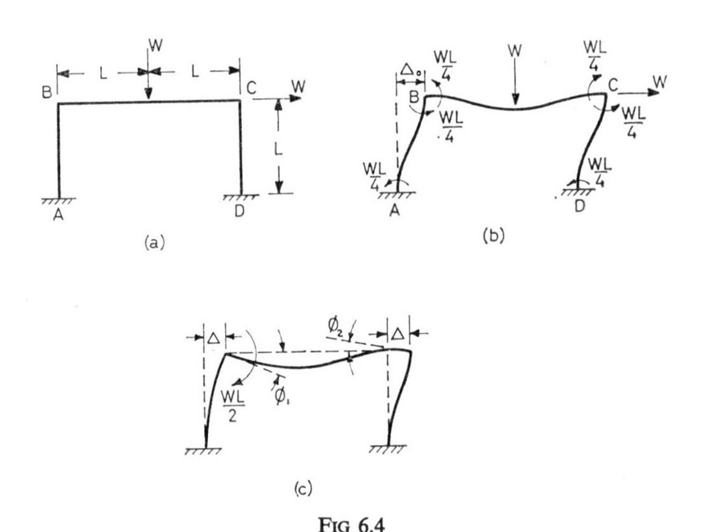

FIG 6.4

at B and C are the fixed end moments, so that there would be no rotations of the joints B and C. The moments $WL/4$ at each end of the columns AB and CD correspond to a sway movement Δ_0 of the frame with no rotations at either end of each column, as illustrated; the corresponding shear force in each column is $W/2$, so that the frame is in horizontal equilibrium. Δ_0 is given by

$$\frac{WL}{4} = \frac{6EI\Delta_0}{L^2},$$

so that

$$\Delta_0 = \frac{WL^3}{24EI} \tag{6.31}$$

In fact, the complete bending moment distribution and the deformations for the loading of Fig. 6.4(b) are known from standard solutions.

Thus the problem is reduced to the solution for the loading of Fig. 6.4(c), with a counter-clockwise moment $WL/2$ applied to the joint B. It will be apparent that any beam or frame problem can similarly be reduced to one in which there are no loads on the members or joints, so that the only external action consists of couples applied at the joints; this initial step is, in fact, always taken in a moment distribution analysis.

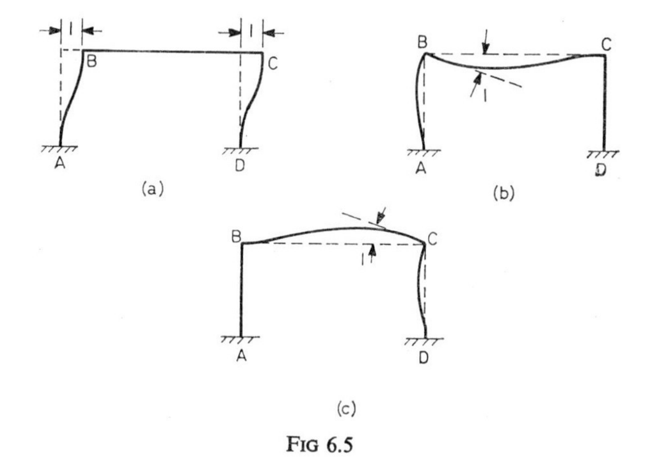

FIG 6.5

The deformation of the whole frame can be expressed in terms of three variables, as indicated in Fig. 6.4(c). These are the joint rotations ϕ_1 and ϕ_2 at B and C, respectively, taken as positive when clockwise, and the sway Δ of BC to the right. The primary objective of the analysis is therefore to determine the values of ϕ_1, ϕ_2 and Δ due to the loading of Fig. 6.4(c).

Three conditions of equilibrium are required for the determination of the three deformation variables. Each of these conditions of equilibrium will be found by means of a virtual work transformation, the actual distribution of bending moment throughout the frame being used in conjunction with three hypothetical deformation systems. These are illustrated in Figs. 6.5(a), (b) and (c), and consist simply of unit values of Δ, ϕ_1, and ϕ_2 in turn, with the other two deformation variables both zero in each case.

The virtual work equation reduces to the form

$$\sum_{\text{joints}} C\phi^{**} = \oint M\kappa^{**}\,\mathrm{d}s, \tag{6.32}$$

where each term $C\phi^{**}$ represents virtual work done by an applied couple C at a joint, and the only difficulty which arises is the evaluation of $\int M\kappa^{**}\,\mathrm{d}s$ for each member. Before proceeding with the particular problem under consideration, some general results which enable the value of this integral to be written down immediately for a member will accordingly be established.

Values of $\int M\kappa^{**}\,\mathrm{d}s$

Figure 6.6(a) shows a uniform prismatic member AB of flexural rigidity EI and length L which has undergone clockwise rotations ϕ_A and ϕ_B at the ends A and B, respectively, together with a sway movement δ, as shown. The end moments, regarded as positive when acting clockwise on the member, are M_{AB} and M_{BA}, respectively. The member is free from lateral load; under these circumstances the slope-deflection equations give the following values for M_{AB} and M_{BA}:

$$\begin{aligned} M_{AB} &= \frac{6EI}{L}\left[\tfrac{1}{3}(2\phi_A+\phi_B)-\frac{\delta}{L}\right] \\ M_{BA} &= \frac{6EI}{L}\left[\tfrac{1}{3}(\phi_A+2\phi_B)-\frac{\delta}{L}\right] \end{aligned} \tag{6.33}$$

The actual bending moment M at a distance x from A, defined as positive when hogging, is therefore given by

F

$$M = -M_{AB}\left(1-\frac{x}{L}\right)+M_{BA}\frac{x}{L}$$

$$= \frac{EI}{L}\left[\frac{6\delta}{L}\left(1-\frac{2x}{L}\right)-4\phi_A\left(1-\frac{3x}{2L}\right)-2\phi_B\left(1-\frac{3x}{L}\right)\right] \tag{6.34}$$

Equation (6.34) defines the actual bending moment at any section of a uniform prismatic elastic member in terms of

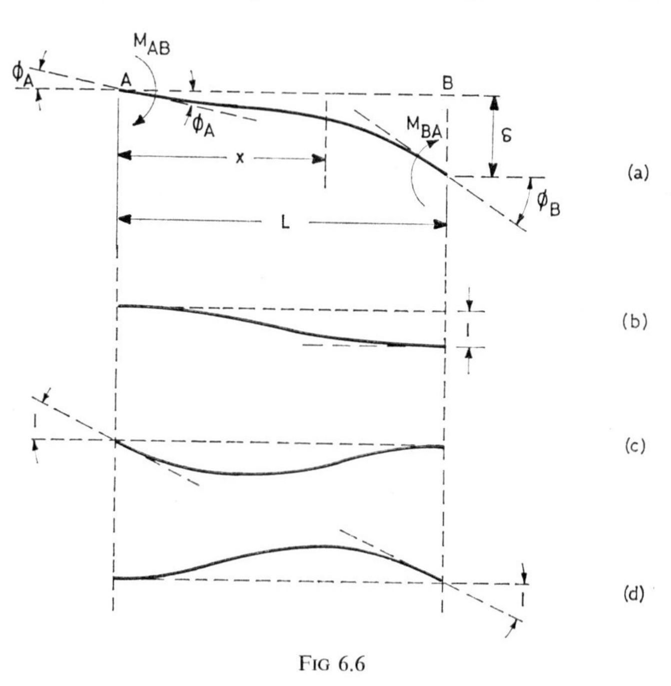

FIG 6.6

deformation variables. Expressions are required for the value of $\int M\kappa^{**}\,ds$ in which M is given by this equation, and κ^{**} represents any one of the three hypothetical curvature distributions depicted in Figs. 6.6(b), (c) and (d).

Taking the unit sway situation of Fig. 6.6(b) first, it is seen from equation (6.34) that κ^{**} is given by the value of M/EI with $\delta = 1$ and $\phi_A = \phi_B = 0$, so that

$$\kappa^{**} = \frac{6}{L^2}\left(1-\frac{2x}{L}\right)$$

Using equation (6.34), it follows that

$$\int M\kappa^{**}\,ds = \int_0^L \frac{EI}{L}\left[\frac{6\delta}{L}\left(1-\frac{2x}{L}\right)-4\phi_A\left(1-\frac{3x}{2L}\right)\right.$$
$$\left.-2\phi_B\left(1-\frac{3x}{L}\right)\right]\frac{6}{L^2}\left(1-\frac{2x}{L}\right)dx$$
$$= \frac{6EI}{L^2}\left[2\frac{\delta}{L}-\phi_A-\phi_B\right]$$

For the unit rotation of Fig. 6.6(c), κ^{**} is given by the value of M/EI in equation (6.34) with $\delta = 0$, $\phi_A = 1$ and $\phi_B = 0$, so that

$$\kappa^{**} = -\frac{4}{L}\left(1-\frac{3x}{2L}\right)$$

Using equation (6.34), it follows that

$$\int M\kappa^{**}\,ds = -\int_0^L \frac{EI}{L}\left[\frac{6\delta}{L}\left(1-\frac{2x}{L}\right)-4\phi_A\left(1-\frac{3x}{2L}\right)\right.$$
$$\left.-2\phi_B\left(1-\frac{3x}{L}\right)\right]\frac{4}{L}\left(1-\frac{3x}{2L}\right)dx$$
$$= -\frac{EI}{L}\left[6\frac{\delta}{L}-4\phi_A-2\phi_B\right]$$

A similar analysis shows that for the unit rotation of Fig. 6.6(d),

$$\int M\kappa^{**}\,ds = -\frac{EI}{L}\left[6\frac{\delta}{L}-2\phi_A-4\phi_B\right]$$

These results are collected together in Table 6.4.

TABLE 6.4

Hypothetical deformation	$\int M\kappa^{**}\, ds$
$\delta = 1$	$\frac{6EI}{L^2}\left[2\frac{\delta}{L}-\phi_A-\phi_B\right]$
$\phi_A = 1$	$-\frac{EI}{L}\left[6\frac{\delta}{L}-4\phi_A-2\phi_B\right]$
$\phi_B = 1$	$-\frac{EI}{L}\left[6\frac{\delta}{L}-2\phi_A-4\phi_B\right]$

Turning now to the problem of Fig. 6.4, a virtual work equation is written down in turn for each of the three hypothetical deformations of Figs. 6.5(a), (b) and (c). The actual deformations for the three members of the frame are first tabulated for reference as follows:

	Actual deformations		
Member	δ	ϕ_A	ϕ_B
AB	Δ	0	ϕ_1
BC	0	ϕ_1	ϕ_2
CD	Δ	ϕ_2	0

Taking Fig. 6.5(a) first, it is seen that there is no rotation at joint B, so that the couple $WL/2$ at this joint does zero virtual work. The virtual work absorbed in the members AB and CD must therefore be zero, and each of these members undergoes a unit sway. It follows from Table 6.4 that

$$\frac{6EI}{L^2}\left[2\frac{\Delta}{L}-0-\phi_1\right]+\frac{6EI}{L^2}\left[2\frac{\Delta}{L}-\phi_2-0\right]=0 \qquad (6.35)$$

For the deformations of Figs. 6.5(b) and (c), the virtual work equations are:

$$-\frac{EI}{L}\left[6\frac{\Delta}{L}-2(0)-4\phi_1\right]-\frac{EI}{2L}[0-4\phi_1-2\phi_2] = \frac{WL}{2} \tag{6.36}$$

$$-\frac{EI}{2L}[0-2\phi_1-4\phi_2]-\frac{EI}{L}\left[\frac{6\Delta}{L}-4\phi_2-0\right] = 0 \tag{6.37}$$

Equations (6.35), (6.36) and (6.37) are the three equations of equilibrium for the determination of Δ, ϕ_1 and ϕ_2. They are found to have the solution

$$\frac{\Delta}{L} = \frac{WL^2}{32EI}, \qquad \phi_1 = \frac{9WL^2}{80EI}, \qquad \phi_2 = \frac{WL^2}{80EI}$$

Using equations (6.33), the bending moments in the frame due to the loading of Fig. 6.4(c) are as given in Table 6.5. Table 6.5 also gives the moments due to the loading of Fig. 6.4(b), together with the sum of these moments which represents the final solution.

TABLE 6.5

	M_{AB}	M_{BA}	M_{BC}	M_{CB}	M_{CD}	M_{DC}
Fig. 6.4(c)	$\frac{3}{80}WL$	$\frac{21}{80}WL$	$\frac{19}{80}WL$	$\frac{11}{80}WL$	$-\frac{11}{80}WL$	$-\frac{13}{80}WL$
Fig. 6.4(b)	$-\frac{1}{4}WL$	$-\frac{1}{4}WL$	$-\frac{1}{4}WL$	$\frac{1}{4}WL$	$-\frac{1}{4}WL$	$-\frac{1}{4}WL$
Total	$-\frac{17}{80}WL$	$\frac{1}{80}WL$	$-\frac{1}{80}WL$	$\frac{31}{80}WL$	$-\frac{31}{80}WL$	$-\frac{33}{80}WL$

As a check on the working, it will be seen in Table 6.5 that the obvious conditions of joint equilibrium, namely

$$M_{BA}+M_{BC} = 0$$

$$M_{CB}+M_{CD} = 0,$$

are satisfied.

The process of moment distribution consists essentially of arriving at the solution of equations (6.35), (6.36) and (6.37) by a trial and error process. Changes are made in an orderly fashion in the values of ϕ_1, ϕ_2 and Δ so that eventually the requirements of equilibrium are all met simultaneously. The moment distribution process is not always recognised as a deformation variable method because the computations are carried out in terms of the bending moments produced by the changes in the deformation variables. Returning to Fig. 6.6, if ϕ_A is changed by an amount ψ while ϕ_B and δ are kept constant, it follows from equations (6.33) that

$$M_{AB} = \frac{4EI}{L}\psi = M, \text{ say}$$

$$M_{BA} = \frac{2EI}{L}\psi = \tfrac{1}{2}M,$$

and it is the values of M and $\frac{1}{2}M$ that are recorded in the calculation.

The equations of equilibrium for this problem could equally well have been established by using the Theorem of Minimum Potential Energy, or equivalently Castigliano's Theorem (Part I). Details of this approach will not be given; the computations involved are identical with those given above.

7

Reciprocal theorems

7.1. INTRODUCTION

In this chapter the Reciprocal Theorems of Maxwell and Betti are derived, and some simple applications of each theorem explained. These theorems are applicable only to linear elastic structures, and are concerned with relationships which exist between the loads which may act on such structures and the deflections which they cause.

Maxwell's Theorem, which may be regarded as a special case of Betti's Theorem, is first proved for a truss system, and then Betti's Theorem is established. In both cases the proof involves merely a simple application of the Principle of Virtual Work. Proofs of the theorems for other types of structure are not given, although there is little difficulty involved in extending the arguments used for trusses to any specified type of structure.

The reciprocal theorems are especially useful for the determination of influence lines. An application of Maxwell's Theorem to the calculation of influence lines for a singly redundant beam is therefore described, and this is followed by an application of Betti's Theorem to the problem of the determination of influence lines for an arch from the results of model tests.

7.2. PROOFS OF RECIPROCAL THEOREMS

Maxwell's Reciprocal Theorem

For the sake of simplicity, the derivation of this theorem will be given for the case of a plane truss. Figure 7.1 shows a plane truss which obeys Hooke's Law. At any two joints A and B, certain directions are prescribed, and the theorem is

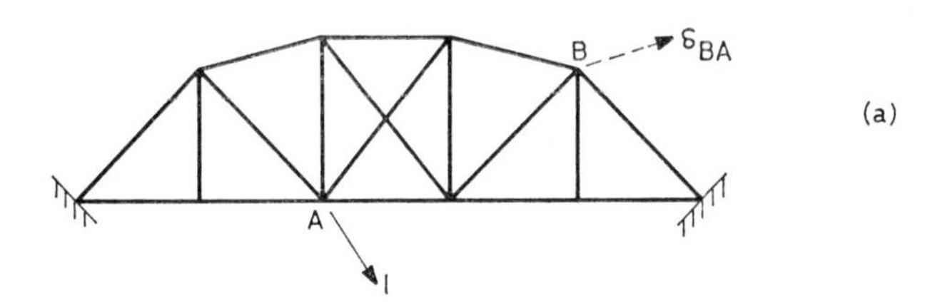

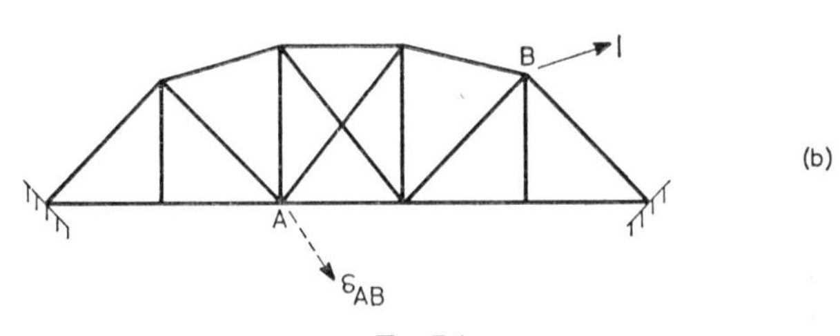

FIG 7.1

concerned with the effects of loads applied at these joints in the prescribed directions, and with the components of deflection which occur in these directions. Thus a unit load at A causes a component of deflection δ_{BA} at B, as shown in Fig. 7.1(a), and a unit load at B causes a component of deflection δ_{AB} at A, as in Fig. 7.1(b). Maxwell's Theorem then states that

$$\delta_{AB} = \delta_{BA}$$

To establish this result, suppose that the unit load at A

causes a force P_A in a typical bar of the truss, so that the extension e_A of this bar is

$$e_A = P_A/\mu, \tag{7.1}$$

where μ is the stiffness of the bar. These bar extensions must be compatible with the actual joint deflections, and in particular with the component of deflection δ_{BA} at B. Suppose also that the unit load at B causes a force P_B in the typical bar of the truss, so that the extension e_B of this bar is

$$e_B = P_B/\mu \tag{7.2}$$

These bar extensions will be compatible with the actual joint deflections, and in particular with the component of deflection δ_{AB} at A.

There are therefore two sets of actual deflections and bar extensions, and two actual force systems available, as follows:

	Forces	*Deformations*
Fig. 7.1(a)	Unit load at A Bar forces P_A	δ_{BA} Bar elongations e_A
Fig. 7.1(b)	Unit load at B Bar forces P_B	δ_{AB} Bar elongations e_B

Two virtual work equations are now written down. The first of these involves using the force system of Fig. 7.1(a) in conjunction with the deformation system of Fig. 7.1(b), and is as follows:

$$1 \times \delta_{AB} = \sum_{\text{bars}} P_A e_B$$

The second equation uses the force system of Fig. 7.1(b) together with the deformation system of Fig. 7.1(a), giving

$$1 \times \delta_{BA} = \sum_{\text{bars}} P_B e_A$$

Using equations (7.1) and (7.2), these equations become

$$\delta_{AB} = \sum_{\text{bars}} P_A P_B/\mu$$

$$\delta_{BA} = \sum_{\text{bars}} P_B P_A/\mu,$$

and it follows at once that

$$\delta_{AB} = \delta_{BA} \tag{7.3}$$

which is Maxwell's Reciprocal Theorem.

This result is applicable to any linear elastic structure, as already stated. It is readily proved by applying the Principle of Virtual Work in the manner given above.

Although Maxwell (1864) was the first to give this result, his work appears to have been unnoticed until Lord Rayleigh (1874) gave a clear statement of its application to specific problems, and also extended the theorem to cover dynamical situations.

Betti's Reciprocal Theorem

This theorem, due to Betti (1872), is an extension of Maxwell's result to cases in which sets of loads and deflections are considered rather than single loads and deflections. Like Maxwell's theorem, it is only applicable to linear elastic structures. It will be proved here for the case of a truss which may be subjected to two sets of loads with three loads in each set, as in Fig. 7.2.

In Fig. 7.2(a) it is supposed that loads W'_A, W'_B and W'_C are applied at A, B and C in the prescribed directions. These loads produce a force P' in a typical bar, giving an extension $e' = P'/\mu$. The components of displacement in the prescribed directions at D, E and F are Δ'_D, Δ'_E and Δ'_F, respectively. Similarly, in Fig. 7.2(b) loads W''_D, W''_E and W''_F are applied at D, E and F, producing a force P'' and extension $e'' = P''/\mu$ in the typical bar. The components of displacement at A, B and C in the prescribed directions are Δ''_A, Δ''_B and Δ''_C.

Betti's Reciprocal Theorem then states that

$$W'_A\Delta''_A + W'_B\Delta''_B + W'_C\Delta''_C = W''_D\Delta'_D + W''_E\Delta'_E + W''_F\Delta'_F \tag{7.4}$$

To establish this result, the Principle of Virtual Work is used in the same way as in the proof of Maxwell's Theorem.

The force and deformation systems which have been defined are

	Forces	*Deformations*
Fig. 7.2(a)	Loads W'_A, W'_B, W'_C Bar forces P'	Δ'_D, Δ'_E, Δ'_F Bar elongations $e' = P'/\mu$
Fig. 7.2(b)	Loads W''_D, W''_E, W''_F Bar forces P''	Δ''_A, Δ''_B, Δ''_C Bar elongations $e'' = P''/\mu$

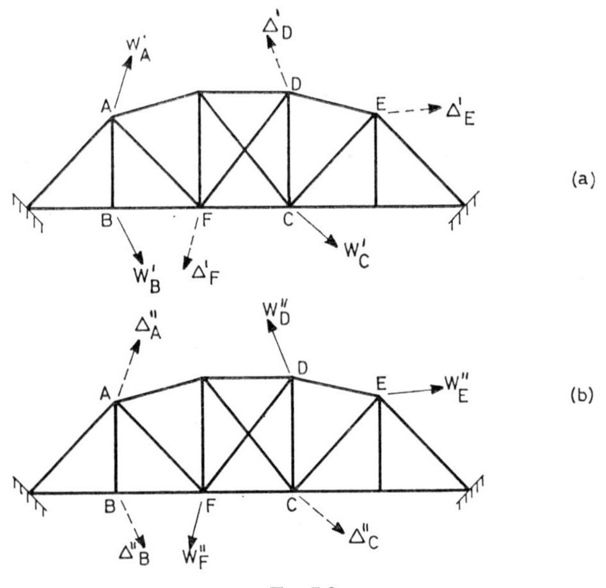

FIG 7.2

Using the force system of Fig. 7.2(a) in conjunction with the displacement system of Fig. 7.2(b), and vice versa, the following two virtual work equations result:

$$W'_A\Delta''_A + W'_B\Delta''_B + W'_C\Delta''_C = \sum_{\text{bars}} P'e'' = \sum_{\text{bars}} P'P''/\mu$$

$$W''_D\Delta'_D + W''_E\Delta'_E + W''_F\Delta'_F = \sum_{\text{bars}} P''e' = \sum_{\text{bars}} P''P'/\mu,$$

and comparison of these two equations shows at once that equation (7.4) is true.

This is the form taken by Betti's Reciprocal Theorem in this particular case; the extension to a larger number of loads in each set presents no difficulty. The theorem also holds true for generalised forces and displacements; for example, in Fig. 7.3,

$$W_1\Delta_1 + W_2\Delta_2 = M_3\phi_3 + M_4\phi_4$$

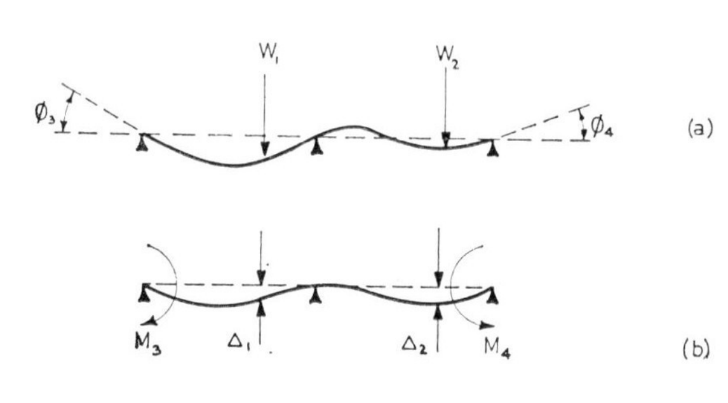

FIG 7.3

7.3. INFLUENCE LINES FOR REDUNDANT STRUCTURES

Maxwell's Reciprocal Theorem can be used to simplify the calculation of influence lines for redundant structures. Consider for example the singly redundant propped cantilever of uniform section with flexural rigidity EI, and span L, which is illustrated in Fig. 7.4(a). The influence line is required for the redundant reaction R_B at the support B.

The value of R_B can be found from the compatibility condition that the vertical deflection at B is zero. If the support at B were removed, as in Fig. 7.4(b), and a unit load was applied at A, the deflection at B may be defined as δ_{BA}. With the support at B still removed, a unit vertical load applied at B would cause a deflection δ_{BB} at B, as in Fig. 7.4(c). By the Principle of

Superposition for linear elastic structures, the deflection at B in the actual situation of Fig. 7.4(a), which is zero by definition, must be given by the compatibility equation

$$0 = \delta_{BA} - R_B \delta_{BB},$$

so that
$$R_B = \frac{\delta_{BA}}{\delta_{BB}} \tag{7.5}$$

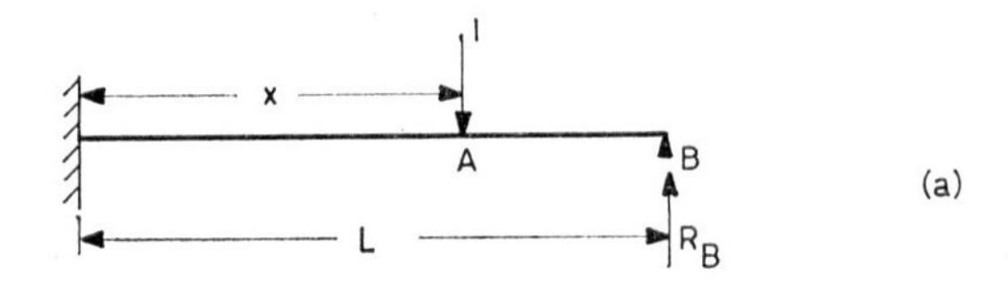

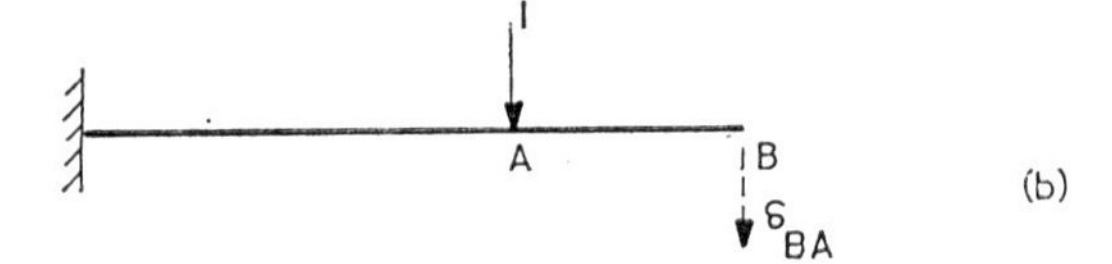

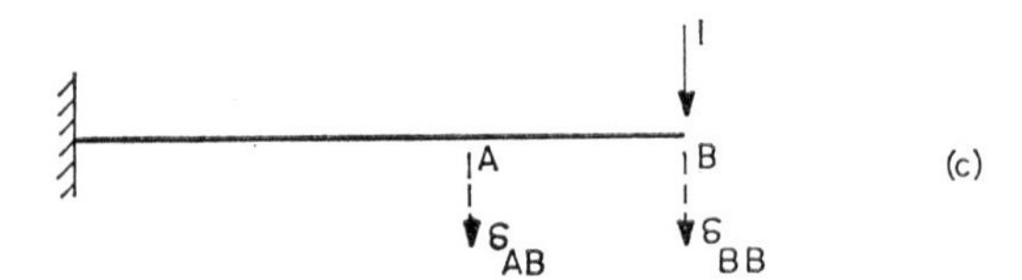

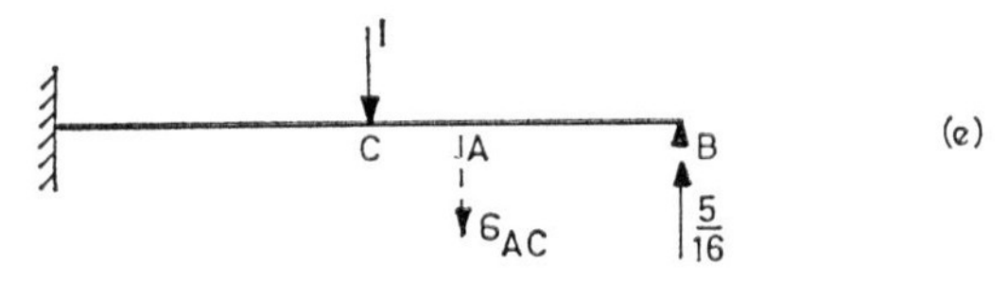

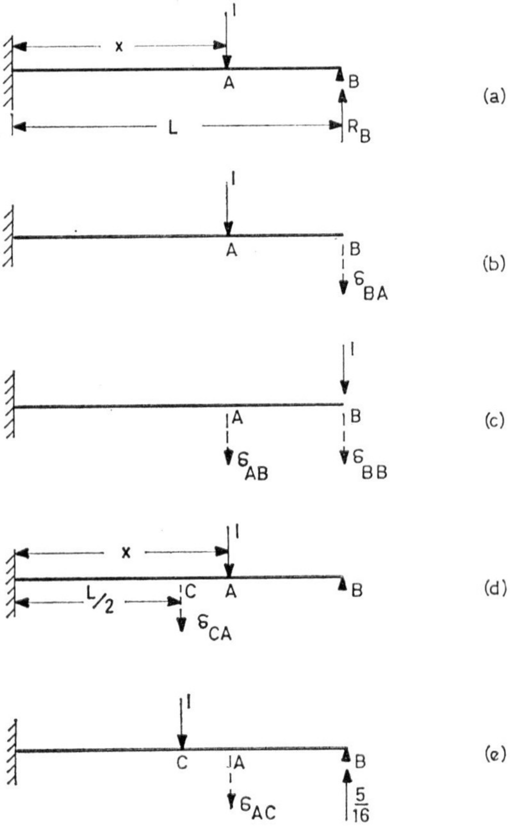

FIG 7.4

Maxwell's Reciprocal Theorem is now applied to the situations depicted in Figs. 7.4(b) and (c), from which it is seen that

$$\delta_{BA} = \delta_{AB}$$

It follows from equation (7.5) that

$$R_B = \frac{\delta_{AB}}{\delta_{BB}} \tag{7.6}$$

The especial merit of this result is that both δ_{AB} and δ_{BB} relate to the statically determinate cantilever situation of Fig. 7.4(c). Thus δ_{BB} is merely the tip deflection of the cantilever carrying a unit concentrated load at the free end, and is therefore $L^3/3EI$. Moreover, δ_{AB} is the deflection of the same cantilever at a distance x from the clamped end, and this is readily shown to be

$$\delta_{AB} = \frac{L^3}{6EI}\left[3\left(\frac{x}{L}\right)^2 - \left(\frac{x}{L}\right)^3\right]$$

It follows from equation (7.6) that

$$R_B = \tfrac{1}{2}\left[3\left(\frac{x}{L}\right)^2 - \left(\frac{x}{L}\right)^3\right] \tag{7.7}$$

The implication of equations (7.6) and (7.7) is that the required influence line is merely the deflection curve for the cantilever of Fig. 7.4(c) drawn to a suitable scale. This is an example of Müller–Breslau's Principle (Müller–Breslau, 1886).

If influence lines for the reaction and moment at the clamped end, or the shear force or bending moment at any section, are also required, it is then a simple matter to derive these from a knowledge of R_B.

If the influence line for the deflection at a particular section C, say at a distance $L/2$ from the clamped end, is required, Maxwell's Reciprocal Theorem can again be used with advantage. Fig. 7.4(d) defines the deflection δ_{CA} at C due to a unit load at A. By Maxwell's theorem $\delta_{CA} = \delta_{AC}$, where δ_{AC}

is the deflection at A due to a unit load at C. It is therefore only necessary to compute δ_{AC} for the situation of Fig. 7.4(e), which has the advantage that the position of the load is now fixed.

The value of δ_{AC} may be found by the "dummy unit load" method described in Chapter 5. From equation (7.7), with $x = L/2$, R_B is found to be $\frac{5}{16}$. The distribution of hogging bending moment M for the situation of Fig. 7.4(e) is therefore

$$M = -\tfrac{5}{16}(L-z) \qquad \text{for } \tfrac{1}{2}L \leqslant z \leqslant L$$

$$M = (\tfrac{1}{2}L-z)-\tfrac{5}{16}(L-z) \qquad \text{for } 0 \leqslant z \leqslant \tfrac{1}{2}L,$$

where z is measured from the clamped end, and the actual curvature κ at any section is M/EI. These curvatures are compatible with the actual deflection δ_{AC} at A, and are used in the equation of virtual work. A force system is used which satisfies the requirements of equilibrium with a hypothetical unit vertical load at A; one such system is obtained by removing the support at B to render the beam statically determinate, giving the following hypothetical bending moment distribution:

$$\text{Hogging moment } M^* = (x-z) \qquad \text{for } 0 \leqslant z \leqslant x$$

$$\text{Hogging moment } M^* = 0 \qquad \text{for } x \leqslant z \leqslant L$$

Using this hypothetical force system in conjunction with the actual displacements and curvatures, the virtual work equation becomes:

$$1.\delta_{AC} = \int_0^L M^*\kappa \, dz = \int_0^x (x-z)\kappa \, dz$$

Thus if $x < \frac{1}{2}L$,

$$\delta_{AC} = \int_0^x (x-z)\frac{1}{EI}\left[(\tfrac{1}{2}L-z)-\tfrac{5}{16}(L-z)\right] dz$$

$$= \frac{1}{96EI}\left[9Lx^2-11x^3\right]$$

while if $x > \frac{1}{2}L$

$$\delta_{AC} = \int_0^{\frac{1}{2}L} (x-z)\frac{1}{EI}\left[(\tfrac{1}{2}L-z)-\tfrac{5}{16}(L-z)\right] dz$$

$$+\int_{\frac{1}{2}L}^{x} (x-z)\frac{1}{EI}\left[-\tfrac{5}{16}(L-z)\right] dz$$

$$= \frac{1}{96EI}\left[5x^3-15Lx^2+12L^2x-2L^3\right]$$

7.4. INFLUENCE LINES BY MODEL ANALYSIS

Influence lines for redundant reactions in elastic structures may be found with the aid of structural models, and Betti's Reciprocal Theorem is then required to establish the basic results.

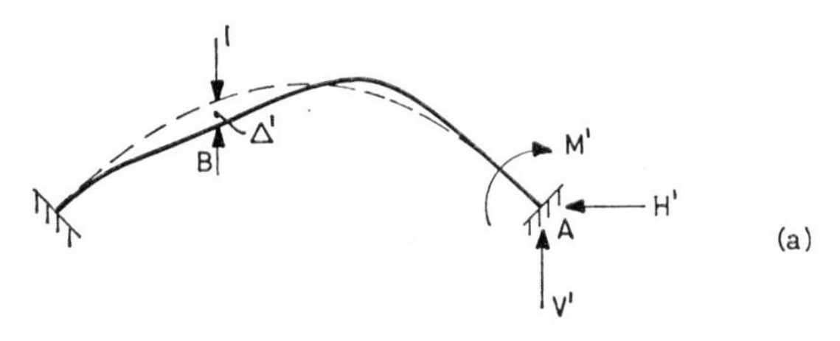

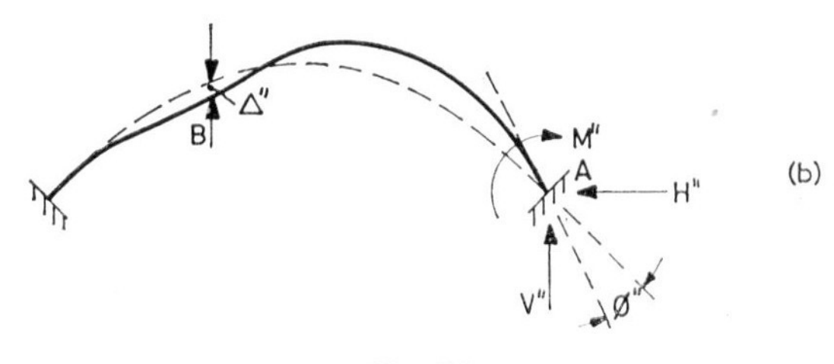

FIG 7.5

Consider for example the arch shown in Fig. 7.5(a), and suppose that the three redundancies are chosen as H, V and M, the horizontal and vertical components of reaction and the hogging moment at A. The problem is to determine the influence lines for each of these three redundancies; the influence line for M will be considered first.

Suppose that a unit vertical load were applied to the arch at some arbitrary section B, producing a hogging moment M' at A together with horizontal and vertical components of reaction H' and V'. The corresponding deflection at B is defined as Δ'.

The experimental procedure which is adopted is to impose a known rotation ϕ'' of the arch at A, while keeping the horizontal and vertical displacements at this abutment zero. The hogging moment required to produce this rotation is defined as M'', as in Fig. 7.5(b), and there will be in addition horizontal and vertical components of reaction H'' and V''. The deflection of the arch at B, which is measured in the experiment, is Δ''.

Two systems of forces and corresponding displacements have now been defined in Figs. 7.5(a) and (b). These are as follows:

		Force	*Corresponding displacement*
Fig. 7.5(a)	B :	1	Δ'
	A :	H'	0
		V'	0
		M'	0
Fig. 7.5(b)	B :	0	Δ''
	A :	H''	0
		V''	0
		M''	ϕ''

Betti's Reciprocal Theorem is now applied, the product of the forces of Fig. 7.5(a) and the displacements of Fig. 7.5(b) being equated to the product of the forces of Fig. 7.5(b) and the displacements of Fig. 7.5(a). This gives

$$\begin{aligned} 1\times\Delta''+H'\times 0+V'\times 0+M'\times\phi'' &= \\ &= 0\times\Delta'+H''\times 0+V''\times 0+M''\times 0, \end{aligned}$$

and it follows immediately that

$$M' = -\frac{\Delta''}{\phi''}$$

Since ϕ'' is known, it follows that M' is proportional to Δ'', so that the required influence line can be found at once by plotting Δ'', which is simply the vertical deflection of the arch caused by the imposition of the rotation ϕ'' at A.

Influence lines for H and V may be determined in a similar fashion. For instance, to find H', a horizontal displacement of A is made while keeping the vertical displacement and rotation at this abutment zero.

Structural models which are used in this way are usually made with a higher flexibility than would result from true scaling, so that the deflections are large and can be measured easily.

8

Theorems of plastic analysis for plane frames

8.1. INTRODUCTION

In this chapter, theorems are discussed which are concerned with plane frames whose members possess the property of forming plastic hinges. The term plane frame here refers to a plane structure, subjected only to loads lying within its plane, which is unbraced and carries these loads primarily by virtue of the resistance of its members to flexure, and for which all the equations of equilibrium relating the bending moments to the applied loads are linear. Simple examples of plane frames are continuous beams and portal frames.

A plastic hinge is defined as a hinge which can only undergo rotation when the bending moment reaches a limiting value M_p, termed the fully plastic moment. For bending moments less than M_p in magnitude, no rotation is possible, but if the bending moment remains constant at the value M_p the rotation at the hinge can increase indefinitely. The sense of the rotation conforms with that of the bending moment, so that work is always absorbed at a plastic hinge.

It will also be assumed throughout this chapter that at a cross-section where a plastic hinge does not form, the relationship between a change in bending moment δM and the corresponding change in curvature $\delta\kappa$ is the elastic relation

$$\delta M = EI\delta\kappa, \tag{8.1}$$

where EI is the flexural rigidity of the cross-section. This assumption is restrictive in that the theorems of plastic analysis can be extended to cover the case in which plastic curvatures may develop in the members at sections adjacent to the plastic hinges (Neal, 1963). However, all the main features of the theorems are brought out if this simpler assumption is retained.

The value of the fully plastic moment for a member is not constant, but depends on a number of factors, notably the shear and axial forces at the plastic hinge position. However, in many practical cases the change in the fully plastic moment produced by these causes is small, and for the purposes of this chapter will be assumed to be negligible.

A full discussion of the practical significance of the plastic hinge concept would be out of place here; it suffices to state that mild steel members in flexure behave in a manner which is in quite close accordance with the concept (Maier–Leibnitz, 1929). Large changes of slope occur in short lengths of such members due to strongly localised plastic flow whenever a limiting value of the bending moment is attained.

For trusses, the counterpart of the plastic hinge property would be that the members would elongate or contract indefinitely whenever a limiting load was reached in tension or compression. Such a hypothesis is in fact obeyed by many tension members, but when compression members buckle, further axial contraction is accompanied by a sharp decrease in the compressive load. Thus the hypothesis is of no practical importance for trusses, which are therefore excluded from further discussion.

The theorems which will be discussed are concerned with the circumstances under which a plane frame may develop indefinitely large deflections due to plastic hinge rotations. There are two ways in which this can happen, one being that a single application of some combination of loads causes the formation of a sufficiently large number of plastic hinges to permit a mechanism motion, in which case plastic collapse is said to occur. The other possibility is that various load combinations may be applied in succession, each of which causes rotations at some plastic hinges; if these eventually build up indefinitely incremental collapse is said to occur.

The theorems of plastic analysis are concerned with the values of the loads which will cause failure by either plastic or incremental collapse. Before these are dealt with in detail, a preliminary discussion of the basic phenomena is given in Section 8.2. The plastic collapse theorems are then stated and proved in Section 8.3, and in Section 8.4 it is shown how these theorems can be used to form upper and lower bounds on the value of the plastic collapse load. The theorems of incremental collapse are then discussed in Section 8.5, and a simple application of these theorems is given in Section 8.6.

There are various methods of analysis for the determination of plastic and incremental collapse loads, each of which is based on one of the theorems of plastic or incremental collapse. No attempt is made to describe all of these methods, which are dealt with in the companion volume by Heyman (1964).

8.2. PLASTIC AND INCREMENTAL COLLAPSE

Plastic collapse

The phenomenon of plastic collapse will be discussed in relation to the rectangular portal frame whose dimensions and loading are as shown in Fig. 8.1(a). This frame is supposed to have rigid joints and to be rigidly built in at its feet, and the fully plastic moment of each member is M_p. The frame is subjected to horizontal and vertical loads which always remain equal, so that the loading is completely specified by the value W of either load. It is then imagined that W is increased steadily until plastic collapse occurs at some value W_c. This is a simple example of the type of loading referred to as proportional loading, in which the loads always bear the same ratio to one another.

There are only five possible plastic hinge positions in this frame, namely A, B, C, D and E. This follows from the fact that in a straight segment of a member which is free from transverse load, such as AB, BC, CD, or DE, the shear force is constant, so that the bending moment varies linearly with distance along the member. Since the bending moment cannot exceed M_p in magnitude, the value M_p can only be attained at one or both ends of such a segment.

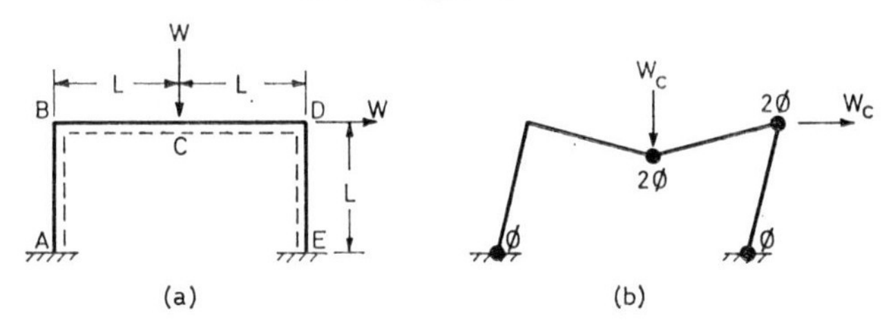

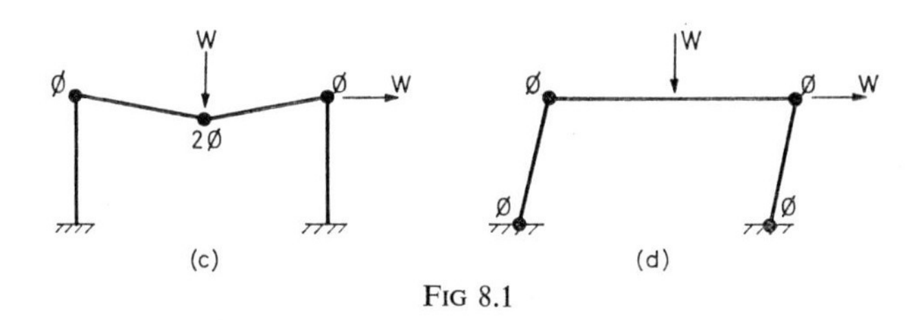

FIG 8.1

The behaviour of the frame as W is increased is summarised in Fig. 8.2, in which W is plotted against h, the horizontal deflection of D. The (W,h) relation is at first wholly elastic, until the bending moment at the most highly stressed section E reaches the fully plastic value M_p. The corresponding load, $2{\cdot}424M_p/L$, is termed the yield load and is denoted by W_y. If W is increased above W_y, the plastic hinge at E rotates while the bending moment at that section remains constant, and the slope of the (W,h) relation is decreased. In succession, plastic hinges form and undergo rotation at sections D and C, and finally at A. The formation of the fourth hinge reduces the structure to a mechanism, as shown in Fig. 8.1(b), and the deflection can then increase indefinitely while W remains constant, the increase of deflection being due solely to motion of this mechanism. This behaviour is termed plastic collapse, and the mechanism of Fig. 8.2(b) is therefore referred to as the plastic collapse mechanism. The corresponding value of W, namely $3M_p/L$, is the plastic collapse load, and is denoted by W_c.

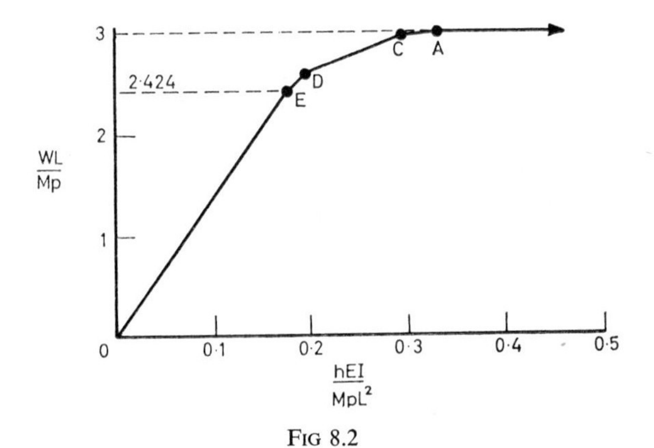

FIG 8.2

The results summarised in Fig. 8.2 are obtained by a step-by-step process of analysis. After the first hinge forms at section E, the response of the structure to an increment of load is the same as for an elastic frame with a pin-joint at this section, since there is no change of bending moment at the plastic hinge. The additional load required to bring the bending moment at D up to the fully plastic value can therefore be computed. For a further increment of load, the structure then responds as though there were pin-joints at E and D, until the bending moment at C reaches the value M_p. Proceeding in this manner, the behaviour up to the plastic collapse load W_c can be determined.

Direct calculation of plastic collapse load

The step-by-step method of analysis could be used for the determination of W_c, but the process would be impossibly laborious for all but the simplest frames. Fortunately, the plastic collapse load can be calculated directly, without considering the behaviour of the frame at lower loads. One way in which this can be done is by writing down a work equation for a small motion of the collapse mechanism. Thus in the mechanism of Fig. 8.1(b), each load moves through a distance $L\phi$, so that the work done by each load is $W_cL\phi$. The work absorbed at each plastic hinge is positive, and is equal to the product of M_p and the magnitude of the hinge rotation. Equating the work done to the work absorbed,

$$W_cL\phi + W_cL\phi = M_p\phi + 2M_p\phi + 2M_p\phi + M_p\phi$$

$$2W_cL\phi = 6M_p\phi$$

$$W_c = 3M_p/L \tag{8.2}$$

This calculation suffices only if it is known that the mechanism used is the actual collapse mechanism. If the results of the step-by-step analysis had not been available, two other possible collapse mechanisms, as shown in Figs. 8.1(c) and (d), would also have had to be considered. For each of these mechanisms a work equation can also be written down, and it is readily shown that in both cases the corresponding value of W is $4M_p/L$.

The choice between these three possible mechanisms of collapse can be made by using the Kinematic Theorem of plastic collapse, which is to be discussed in Section 8.3. In the terminology of this particular example, this theorem states that the plastic collapse load W_c is the smallest value of W which is found to correspond to any possible plastic collapse mechanism. The mechanism of Fig. 8.1(b) is therefore the actual plastic collapse mechanism, and W_c is $3M_p/L$.

In general, the Kinematic Theorem, and the other theorems

of plastic collapse, enable the value of the plastic collapse load to be calculated in various ways. In each method, the calculation is direct, and there is no reference to the previous history of loading and deformation of the frame.

Incremental collapse

The phenomenon of incremental collapse will also be described with reference to the portal frame of Fig. 8.1(a). For this frame the behaviour under proportional loading was indicated in Fig. 8.2; by contrast, its behaviour when subjected to successive cycles of load will now be discussed.

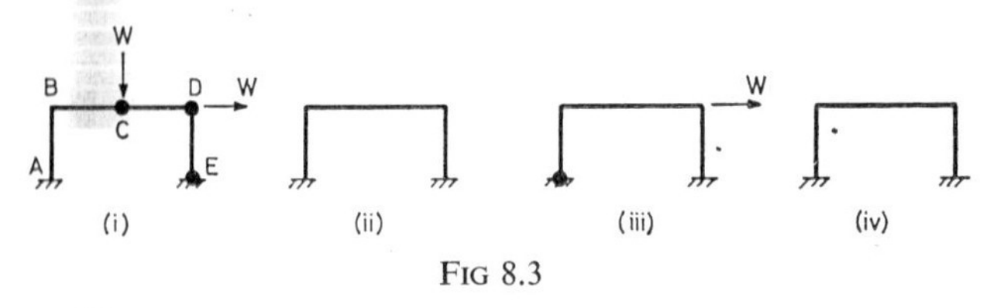

FIG 8.3

A single cycle of loading is defined in Fig. 8.3. This cycle consists of the following four steps:

(i) Application of horizontal and vertical loads, each of magnitude W
(ii) Removal of these loads
(iii) Application of a horizontal load of magnitude W
(iv) Removal of this load.

The effect produced by several cycles of loading depends on the value of W. Figure 8.4 shows the horizontal deflection h when both loads are applied simultaneously, as in Fig. 8.3(i), as a function of n, where n is the number of applications of these two loads.

Until W exceeds $2{\cdot}737M_p/L$, rotations of plastic hinges only occur, if at all, during the first loading, and all subsequent changes of load are borne by wholly elastic changes of bending moment. There is therefore no growth of deflection with number of cycles. However, when W exceeds $2{\cdot}737M_p/L$, the deflection increases with number of cycles. If W is less than $2{\cdot}857M_p/L$, the total increase of deflection is limited, since as n tends to infinity the deflection tends to a finite limit. The frame ultimately attains a condition in which all subsequent changes of load are borne by purely elastic changes of bending moment, and is then said to have shaken down. For values of W greater than $2{\cdot}857M_p/L$, as exemplified for the results for $2{\cdot}90M_p/L$,

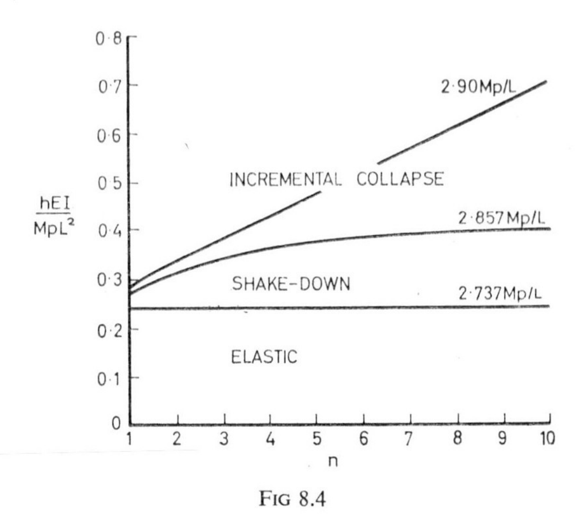

FIG 8.4

there is a constant increment of deflection per cycle, so that indefinitely large deflections are developed. This behaviour is termed incremental collapse.

The step-by-step calculations leading to Fig. 8.4 are similar in principle to those used in deriving Fig. 8.2. Consider for example the calculations for $W = 2{\cdot}90M_p/L$. The behaviour on first loading, with both loads equal as in Fig. 8.3(i), is obtained in precisely the same way as before, and when $W = 2{\cdot}90M_p/L$, plastic hinges have formed and undergone rotation at E and D, as can be seen from Fig. 8.2. When these loads are removed, as in Fig. 8.3(ii), the entire frame behaves elastically. The residual bending moments remaining after their removal are therefore found by subtracting the elastic bending moments which would be caused by the application of these two loads from the actual bending moments which were developed. When the horizontal load only is applied, as in Fig. 8.3(iii), the frame at first responds elastically, and the elastic changes of bending moment are added to the residual bending moments. Eventually, the fully plastic moment is reached at A, and as the horizontal load is increased further to the value $2{\cdot}90M_p/L$, plastic hinge rotation takes place at this cross-section. Removal of the horizontal load, as in Fig. 8.3(iv), then leaves the frame with a different distribution of residual bending moments from that corresponding to the situation of Fig. 8.3(ii).

Further cycles of loading are dealt with in a similar manner. It is found that the third and all subsequent applications of both loads together produces rotation of a plastic hinge at C as well as at D and E. Moreover, the magnitude of the increment in each of the four hinge rotations, at A, C, D and E, is the same in each cycle after the second. These hinge rotations, if they all occurred simultaneously, would constitute the mechanism motion of Fig. 8.1(b), and although they actually occur at different stages during the cycle, are such as to cause a constant increment of deflection per cycle, as shown in Fig. 8.4. Thus deflections of any magnitude can be built up if n is large enough, causing failure by incremental collapse.

For this example it is found that incremental collapse will occur if W exceeds $2{\cdot}857M_p/L$. This limiting value of W is termed the incremental collapse load, and is denoted by W_S.

For values of W between $2{\cdot}737M_p/L$ and $2{\cdot}857M_p/L$, the behaviour is similar, with the important exception that there is no plastic hinge rotation at C, no matter how many cycles of loading are applied. The hinge rotations at A, D and E which occur in each cycle diminish as n increases, and tend to zero as n tends to infinity; moreover, their total magnitude remains finite. Thus the frame ultimately shakes down by reaching a state of residual stress which enables all further variations of load to be borne by purely elastic responses of the frame.

When W is $2{\cdot}857M_p/L$, it is found that after shake-down has occurred, the maximum bending moment at C, which occurs when both loads are applied simultaneously, is equal to M_p, this value being just attained as each load reaches its maximum value. Under these circumstances, there is, of course, no plastic hinge rotation at this section.

Direct calculation of incremental collapse load

The step-by-step calculations are extremely tedious, but fortunately it is possible to calculate the value of W_s by direct methods. These methods are closely analogous to those which are used for the calculation of plastic collapse loads. For instance, W_s can be calculated by considering the kinematics of the incremental collapse mechanism, which is as shown in Fig. 8.1(b).

In performing this calculation, the residual bending moments m which must exist in the structure when it has shaken down after the application of a large number of cycles of load of intensity W_s are first determined. In evaluating these moments, it is necessary to specify a sign convention, and in what follows bending moments will be regarded as positive if they cause tension in those fibres of the member adjacent to the dotted line in Fig. 8.1(a).

Consider for example section E. When both loads W_s are applied simultaneously, the bending moment at this section will be M_p. During removal of these loads, the entire frame behaves elastically, and the change of bending moment, as determined by any of the standard methods of elastic analysis, is $-0{\cdot}412W_sL$. Thus the residual bending moment m_E is given by

$$m_E = M_p - 0{\cdot}412W_sL$$

The residual bending moments at C and D are similarly found to be

$$m_{\mathrm{D}} = -M_p + 0{\cdot}388 W_s L$$
$$m_{\mathrm{C}} = M_p - 0{\cdot}300 W_s L,$$

and at section A, where the bending moment has the value $-M_p$ when the horizontal load only is applied,

$$m_{\mathrm{A}} = -M + 0{\cdot}313 W_s L$$

The residual bending moments satisfy the requirements of equilibrium with zero external loads, and will be used in the virtual work equation. This equation was given in Chapter 3, equation (3.6), as

$$\sum_{\text{joints}} (H^* h^{**} + V^* v^{**} + C^* \phi^{**}) + \oint (w_n^* y_n^{**} + w_t^* y_t^{**})\, \mathrm{d}s$$
$$= \sum_{\text{hinges}} M^* \theta^{**} + \oint (M^* \kappa^{**} + P^* \varepsilon^{**})\, \mathrm{d}s$$

For a bending moment distribution satisfying the requirements of equilibrium with zero external loads, the left-hand side of this equation becomes zero. If in addition the axial strain ε^{**} is everywhere zero, the term $P^*\varepsilon^{**}$ vanishes, and the equation reduces to

$$\sum_{\text{hinges}} M^* \theta^{**} + \oint M^* \kappa^{**}\, \mathrm{d}s = 0 \qquad (8.3)$$

The deformations which will now be used are those of the incremental collapse mechanism of Fig. 8.1(b), for which κ^{**} is everywhere zero, as in any mechanism motion. Thus equation (8.3), with $M^* = m$, and θ^{**} replaced by the actual hinge rotation θ, reduces to

$$\sum_{\text{hinges}} m\theta = 0 \qquad (8.4)$$

Using the actual hinge rotations, whose magnitudes are shown in Fig. 8.1(b), this equation becomes

$$m_{\mathrm{A}}(-\phi) + m_{\mathrm{C}}(2\phi) + m_{\mathrm{D}}(-2\phi) + m_{\mathrm{E}}(\phi) = 0,$$

so that

$$-m_{\mathrm{A}} + 2m_{\mathrm{C}} - 2m_{\mathrm{D}} + m_{\mathrm{E}} = 0 \qquad (8.5)$$

Substituting the values of the residual bending moments just derived,

$$-(-M_p+0{\cdot}313W_sL)+2(M_p-0{\cdot}300W_sL)$$
$$-2(-M_p+0{\cdot}388W_sL)+(M_p-0{\cdot}412W_sL) = 0,$$

whence $$W_s = 2{\cdot}857M_p/L$$

It will be seen that signs are attributed to the hinge rotations which conform with the sign convention for bending moments. Thus at A, the hinge rotation ϕ would cause compression in the fibres adjacent to the dotted line in Fig. 8.1(a), and must therefore be regarded as negative.

Calculations of this kind could also be used to determine values of W corresponding to assumed incremental collapse mechanisms, such as those of Figs. 8.1(c) and (d), as will be seen in Section 8.6. The Kinematic Theorem of incremental collapse, which is discussed in Section 8.5, then permits the correct incremental collapse mechanism to be selected as the one to which there corresponds the lowest value of W.

The Kinematic Theorem, and the other theorems of incremental collapse, enable the value of the incremental collapse load to be calculated by various methods, each of which is analogous to a method for calculating plastic collapse loads. Each method of calculation is direct, and as will be seen in Section 8.5 it is not necessary to assume a particular cycle of loading. All that needs to be known about each load is the upper and lower limits between which it can vary; the variation of each load between its prescribed limits can be quite independent of the nature of the variations of all the other loads. This type of loading is referred to as variable repeated loading.

8.3. PLASTIC COLLAPSE THEOREMS

There are three theorems concerning the value of the plastic collapse load for a given frame and loading, namely the Static, Kinematic and Uniqueness theorems. Before these are discussed, an essential preliminary is to show that during plastic collapse the bending moment distribution remains unaltered, and that the increases of deflection are due solely to rotations at the plastic hinges.

Constancy of bending moments during plastic collapse

Consider a frame which is undergoing plastic collapse under constant loads. During a definite small time interval, let the bending moment at any cross-section change by δM. Since the loads are constant during plastic collapse, the changes of bending moment δM must satisfy the conditions of equilibrium with zero external loads, and can therefore be used in the virtual work equation (8.3).

During the same time let the curvature at a typical cross-section change by $\delta\kappa$ and the rotation at a typical plastic hinge change by $\delta\theta$. These changes of deformation must satisfy the requirements of compatibility with zero axial strains in the members, and can therefore also be used in equation (8.3), which then becomes

$$\sum_{\text{hinges}} \delta M \delta\theta + \oint \delta M \delta\kappa \, \mathrm{d}s = 0 \qquad (8.6)$$

At each plastic hinge position, δM must be zero, since rotation at a plastic hinge can only occur if the bending moment remains constant at the fully plastic value. Equation (8.6) therefore reduces to

$$\oint \delta M \delta\kappa \, \mathrm{d}s = 0 \qquad (8.7)$$

Since the frame remains elastic at every section other than a hinge position, δM and $\delta\kappa$ are related by equation (8.1), namely

$$\delta M = EI\delta\kappa$$

Equation (8.7) therefore becomes

$$\oint \frac{(\delta M)^2}{EI} \, \mathrm{d}s = 0 \qquad (8.8)$$

This equation can only be satisfied if δM is zero at every section, which establishes the result.

From equation (8.1), it follows at once that $\delta\kappa$ is also constant everywhere during plastic collapse, so that the growth of deflection during plastic collapse is due solely to rotations at the plastic hinges. This result was first established by Greenberg (1949), using the terminology of truss-type structures.

Static Theorem of Plastic Collapse

Suppose that a frame is subjected to a set of proportional loads, which bear constant ratios to one another. The set of loads is then completely specified by the value of any one of the loads, say W. Any change in the loads can then be expressed as the load factor λ by which W, and therefore each other load, is multiplied.

The value of λ at which plastic collapse would occur is then defined as λ_c and is termed the collapse load factor.

The Static Theorem of plastic collapse is concerned with distributions of bending moment throughout a frame which satisfy all the requirements of equilibrium with the set of proportional loads λW. Any such distribution will be termed *statically admissible* with these loads. If in addition the bending moment distribution is such that the fully plastic moment is not exceeded anywhere in the frame, the distribution is also referred to as *safe*. The Static Theorem can then be stated as follows:

"If there exists any distribution of bending moments throughout a frame which is statically admissible with a set of proportional loads λW, and also safe, the load factor λ cannot be greater than the plastic collapse load factor λ_c."

To establish this result, let M' denote the bending moment at a typical cross-section in a distribution which is both safe and statically admissible with the set of loads λW. Because the equations of equilibrium are linear, it follows that the bending moment distribution $\lambda_c M'/\lambda$ must be statically admissible with the set of collapse loads $\lambda_c W$. If the actual bending moment distribution at collapse is denoted by M, these bending moments must also be statically admissible with the set of collapse loads $\lambda_c W$. It follows that the bending moment distribution $(M - \lambda_c M'/\lambda)$ must satisfy the requirements of equilibrium with zero external loads, and can therefore be used in the virtual work equation (8.3).

Let $\delta\theta$ denote the rotation of a typical plastic hinge during a small motion of the actual collapse mechanism. The hinge rotations $\delta\theta$ satisfy the requirements of compatibility with zero change of curvature at every section, and can therefore be used in the virtual work equation (8.3), with κ^{**} identically zero.

The virtual work equation therefore becomes

$$\sum_{\text{hinges}} (M - \lambda_c M'/\lambda)\delta\theta = 0,$$

so that

$$\sum_{\text{hinges}} M\delta\theta = \frac{\lambda_c}{\lambda} \sum_{\text{hinges}} M'\delta\theta \tag{8.9}$$

At a hinge position where $\delta\theta$ is positive, the actual bending moment M must be equal to the fully plastic moment M_p. Since the distribution M' was defined as safe, M' cannot exceed M_p. It follows that if $\delta\theta > 0$,

$$M = M_p, \qquad M' \leqslant M_p$$

so that

$$M\delta\theta \geqslant M'\delta\theta$$

On the other hand, if $\delta\theta < 0$,

$$M = -M_p, \qquad M' \geqslant -M_p,$$

so that again

$$M\delta\theta \geqslant M'\delta\theta$$

Thus in equation (8.9), the value of $M'\delta\theta$ at each hinge position cannot exceed the value of $M\delta\theta$. It follows at once that

$$\lambda \leqslant \lambda_c,$$

which establishes the theorem. This result was first stated by Kist (1917) as an axiom; proofs were later supplied independently by Gvozdev (1936), Greenberg and Prager (1949) and Horne (1950).

Feinberg's axiom

An interesting corollary of the Static Theorem was put forward by Feinberg (1948) as an axiom. This is concerned with the effect of strengthening a frame by increasing the fully plastic moment of one or more of its members. Feinberg's axiom states that the collapse load factor cannot thereby be reduced.

If the original frame would collapse under a set of loads $\lambda_c W$, there must be at least one set of bending moments M which is statically admissible with these loads and also safe. These bending moments would continue to be statically admissible with the loads $\lambda_c W$ on the strengthened frame, and would certainly continue to be safe, since the fully plastic moment is nowhere decreased. It follows at once from the Static Theorem that the collapse load factor for the strengthened frame cannot be less than λ_c.

Kinematic Theorem of Plastic Collapse

It was shown in Section 8.2 that a work equation can be written down for any assumed mechanism, which might or might not be the actual collapse mechanism, so that a corresponding value of the load factor λ can be derived. The Kinematic Theorem is concerned with such corresponding values of λ, and is as follows:

"The load factor λ which is found to correspond to any assumed plastic collapse mechanism cannot be less than the plastic collapse load factor λ_c."

This theorem can be proved directly (Gvozdev, 1936), but it can also be derived from Feinberg's axiom (Greenberg and Prager, 1949). Any assumed mechanism of collapse for a given frame and loading, giving rise to a corresponding load factor λ, would be the actual collapse mechanism for this frame if the fully plastic moment was increased indefinitely at all cross-sections except those at which plastic hinges occur in the assumed mechanism, where the fully plastic moments are left unchanged. From Feinberg's axiom, this strengthening could not reduce the collapse load factor, so that λ_c for the original frame could not exceed λ.

Uniqueness Theorem of Plastic Collapse

This theorem is as follows:

"If there exists at least one bending moment distribution throughout a frame which is both safe and statically admissible with a set of proportional loads λW, and in which the fully plastic moment occurs at enough cross-sections to cause failure of the frame as a mechanism, the load factor λ must be equal to the plastic collapse load factor λ_c."

The first set of requirements stipulated in this theorem are precisely those of the Static Theorem, so that λ cannot exceed λ_c. The other requirements are those of the Kinematic Theorem, and λ cannot therefore be less than λ_c. It follows that λ must be equal to λ_c when all these requirements are met simultaneously.

The theorem may thus be regarded as a combination of the Static and Kinematic Theorems; a separate proof was given by Horne (1950).

8.4. UPPER AND LOWER BOUNDS ON λ_c

Various methods for determining the collapse load factor λ_c have been developed, each one being based on one of the plastic collapse theorems described in Section 8.3. Descriptions of these methods will not be given here; the reader is referred instead to accounts such as those of Heyman (1964) or Neal (1963). Instead, attention will here be confined to a simple application of the Static and Kinematic Theorems to form upper and lower bounds on the value of λ_c.

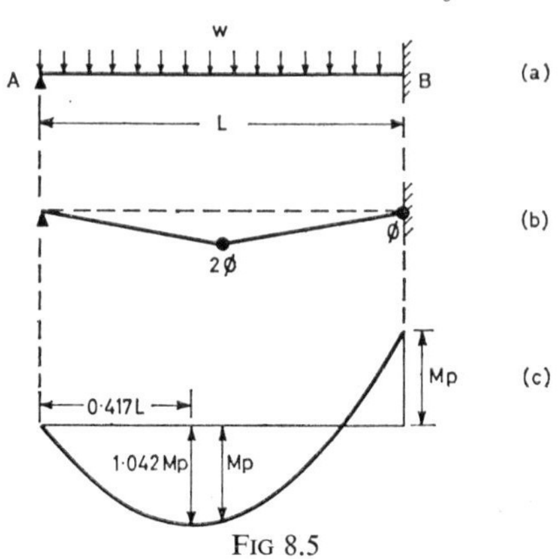

FIG 8.5

Consider the propped cantilever of length L, as shown in Fig. 8.5(a), which carries a uniformly distributed load of intensity w. The fully plastic moment is M_p, and it is required to find the collapse load factor λ_c.

Plastic collapse would evidently occur with the formation of two plastic hinges, one at the fixed end B and one at some position within the span. Because the load is uniformly distributed, the position of this latter hinge is not known *a priori*; it will actually occur at the position of maximum sagging bending moment.

Suppose that a plastic hinge is first assumed to occur at mid-span, as in Fig. 8.5(b). The deflection at the hinge in the mechanism motion shown is $0{\cdot}5L\phi$, so that the average deflection of the uniformly distributed load is $0{\cdot}25L\phi$. Equating the work done to the work absorbed, it is found that

$$(\lambda_1 wL)0{\cdot}25L\phi = 3M_p\phi$$

$$\lambda_1 = 12M_p/wL^2, \qquad (8.10)$$

λ_1 being the load factor corresponding to the assumed mechanism.

From the Kinematic Theorem, this load factor cannot be less than the collapse load factor λ_c, so that

$$\lambda_c \leqslant 12M_p/wL^2 \tag{8.11}$$

This calculation thus gives an upper bound on the value of λ_c. A lower bound can also be obtained from the same mechanism by considerations of statics. Fig. 8.5(c) shows the bending moment diagram corresponding to the assumed mechanism; this is readily derived from a knowledge of the load intensity $\lambda_1 w$, which is $12M_p/L^2$, from equation (8.10). It will be seen that the greatest sagging bending moment in the beam is $1{\cdot}042M_p$, occurring at a distance $0{\cdot}417L$ from A.

Since the equations of equilibrium are linear, it follows at once that if the bending moments were all reduced in the same ratio, 1 : 1·042, the distribution thus obtained would be statically admissible with a distributed load whose intensity corresponded to the load factor $\lambda_1/1{\cdot}042$. This distribution would also be safe, since the greatest bending moment would then be of magnitude M_p. It follows from the Static Theorem that this reduced load factor could not be greater than the actual collapse load factor λ_c, so that the following lower bound on the value of λ_c is obtained:

$$\frac{\lambda_1}{1{\cdot}042} = 11{\cdot}52M_p/wL^2 \leqslant \lambda_c \tag{8.12}$$

Combining this result with (8.11),

$$11{\cdot}52M_p/wL^2 \leqslant \lambda_c \leqslant 12M_p/wL^2 \tag{8.13}$$

If these bounds were not close enough for practical purposes, a fresh calculation could be made for a mechanism with a hinge at a distance $0{\cdot}417L$ from A, where the maximum bending moment occurs in the distribution of Fig. 8.5(c). This calculation gives upper and lower bounds which are both equal to $11{\cdot}66M_p/wL^2$ to four significant figures.

It is not difficult to determine the precise position of the plastic hinge within the span for this problem, and hence the actual value of λ_c. One procedure is to write down the work equation with this hinge at a variable distance x from A. This gives λ as a function of x, and the correct value of x is the value which minimises λ, by the Kinematic Theorem. It is found that this value of x is $(\sqrt{2}-1)L$, and the value of λ_c is $(6+4\sqrt{2})M_p/wL^2$, or $11{\cdot}66M_p/wL^2$.

8.5. SHAKE-DOWN AND INCREMENTAL COLLAPSE THEOREMS

As in the case of plastic collapse, there are three theorems concerning the value of the load factor λ_s above which failure by incremental collapse would occur for a frame subjected to variable repeated loading. These theorems are closely analogous to the plastic collapse theorems, and are termed the Static, Kinematic and Uniqueness Theorems of incremental collapse.

The Static Theorem is part of a general theorem, known as the Shake-down Theorem, which states the conditions under which a frame which is subjected to variable repeated loading will ultimately shake down. This latter theorem will first be proved, and the Static Theorem will then be stated. Finally, derivations of the Kinematic and Uniqueness Theorems will be given.

Variable repeated loading may be defined as follows. Suppose that any particular load W on a frame can never exceed a maximum W^{max} or fall below a minimum W^{min}, so that

$$W^{max} \geqslant W \geqslant W^{min} \tag{8.14}$$

Subject only to this restriction, each load can vary independently of the values of the other loads in any manner whatsoever. This type of loading is of course more general than the cyclic loading which was considered in Section 8.2 for the purpose of illustrating the incremental collapse phenomenon.

The set of loads defined by inequalities such as (8.14) will be referred to collectively as the variable repeated loads W. If each load limit is multiplied by a load factor λ, the set of loads may be referred to as λW.

If the frame remained entirely elastic when subjected to the set of loads W, then at any typical cross-section the bending moment $\mathscr{M}$ would be linearly proportional to the value of each load. From the conditions (8.14) the maximum and minimum possible values of this elastic bending moment, $\mathscr{M}^{\max}$ and $\mathscr{M}^{\min}$, could therefore be deduced. If the value of each load limit was increased by a load factor λ, these maximum and minimum elastic moments would become $\lambda\mathscr{M}^{\max}$ and $\lambda\mathscr{M}^{\min}$.

The Shake-down Theorem can now be stated as follows:

"If a frame is subjected to a set of variable repeated loads λW, and there exists any distribution of residual bending moments $\overline{m}$ which satisfies at every cross-section the inequalities

$$\left.\begin{aligned} \overline{m}+\lambda\mathscr{M}^{\max} &\leqslant M_p \\ \overline{m}+\lambda\mathscr{M}^{\min} &\geqslant -M_p \end{aligned}\right\} \tag{8.15}$$

the frame will eventually shake down."

In this statement of the theorem, the residual bending moments $\overline{m}$ are presumed to satisfy all the requirements of equilibrium with zero external loads.

The theorem is proved by considering the behaviour of a quantity Q defined by

$$Q = \oint \frac{(m-\overline{m})^2}{2EI}\,\mathrm{d}s, \tag{8.16}$$

where m is the actual residual bending moment at any typical cross-section at any stage during the loading. m is defined by

$$m = M-\lambda\mathscr{M}, \tag{8.17}$$

where M is the actual bending moment at any instant, and $\lambda\mathscr{M}$ is the elastic bending moment due to the same loading. The residual bending moment m, defined in this way, would only be the actual bending moment in the unloaded frame if the entire frame behaved elastically during unloading.

The bending moment distributions M and $\lambda\mathscr{M}$ both satisfy the requirements of equilibrium with the same external loads, and it follows that the residual moments m, defined by equation (8.17), must satisfy the requirements of equilibrium with zero external loads.

From the definition of (8.16), it follows that Q is always positive; it may be regarded as a measure of the difference between the actual and hypothetical residual moment distributions m and $\overline{m}$.

The proof consists of showing that whenever there is any plastic hinge rotation, Q decreases; it must therefore ultimately become zero, in which case m becomes equal to $\overline{m}$ everywhere, or else settle down at some positive value and thereafter remain unchanged. In either case, the frame would then have shaken down.

During a definite small time interval, suppose that small changes occur in the applied loads, such that at a typical cross-section the actual bending moment changes by δM, with corresponding changes $\lambda\delta\mathscr{M}$ and δm in the elastic and residual bending moments. It follows from equation (8.16) that the corresponding change in Q is given by

$$\delta Q = \oint (m-\overline{m}) \frac{\delta m}{EI}\, ds$$

$$= \oint (m-\overline{m}) \frac{(\delta M - \lambda\delta\mathscr{M})}{EI}\, ds, \qquad (8.18)$$

using equation (8.17).

Equation (8.18) is now transformed by using the virtual work equation. The actual changes of curvature $\delta M/EI$ must be compatible with the actual changes $\delta\theta$ in the rotations at the plastic hinges which occur during the same interval. Furthermore, the elastic changes of curvature $\lambda\delta\mathscr{M}/EI$ must be compatible with zero plastic hinge rotations. Thus the changes of curvature $(\delta M - \lambda\delta\mathscr{M})EI$ must be compatible with the changes in plastic hinge rotation. Using these deformations in the equation of virtual work (8.3), in conjunction with bending moments $(m - \overline{m})$, which must satisfy the conditions of equilibrium with zero external loads,

$$\sum_{\text{hinges}} (m-\overline{m})\delta\theta + \oint (m-\overline{m}) \frac{(\delta M - \lambda\delta\mathscr{M})}{EI} \, ds = 0 \quad (8.19)$$

Combining this with equation (8.18), it is seen that

$$\delta Q = - \sum_{\text{hinges}} (m-\overline{m})\delta\theta \quad (8.20)$$

Suppose that a particular hinge position,

$$m < \overline{m}$$

It follows from the first of the inequalities (8.15) that

$$m + \lambda\mathscr{M}^{\max} < M_p \quad (8.21)$$

From equation (8.17), the actual bending moment M at this position is

$$M = m + \lambda\mathscr{M},$$

so that from equation (8.21),

$$M < M_p$$

Thus the rotation $\delta\theta$ at this hinge position must be negative, and therefore

$$(m-\overline{m})\delta\theta > 0 \quad (8.22)$$

It can be shown similarly that if $m > \overline{m}$, $\delta\theta$ must be positive, so that again $(m-\overline{m})\delta\theta$ is positive.

Combining the condition (8.22) with equation (8.20), it is seen that

$$\delta Q < 0$$

If there are no plastic hinge rotations during the interval considered, δQ must be zero, from equation (8.20). It follows that

Q decreases whenever any plastic hinge rotation occurs, but remains constant if the frame behaves elastically. As already stated, this establishes the Shake-down Theorem.

The Shake-down Theorem was first proved by Bleich (1932) for structures with not more than two redundancies. A general proof, using the terminology of truss-type structures, was first given by Melan (1936). This was later simplified by Symonds and Prager (1950), and an adaptation to frames was given by Neal (1951).

The above proof does not discuss the deflections which may have developed by the time shake-down has occurred. However, it can be shown (Neal, 1963), that these will be finite.

Incremental collapse and alternating plasticity

If the conditions for shake-down to occur, as expressed by the inequalities (8.15), are not met, plastic hinge rotations can continue to build up indefinitely. This can cause failure by incremental collapse, as explained in Section 8.2; alternatively, it is possible that at one particular section alternate positive and negative rotations of a hinge may occur in sequence without limit, thus leading eventually to fracture. This type of failure is termed alternating plasticity; the load factor λ_a at which it will occur is determined by expressing (8.15) as the continued inequality

$$-M_p-\lambda\mathscr{M}^{\min}\leqslant\overline{m}\leqslant M_p-\lambda\mathscr{M}^{\max},$$

so that

$$\lambda(\mathscr{M}^{\max}-\mathscr{M}^{\min})\leqslant 2M_p$$

This is evidently the condition for the avoidance of alternating plasticity at a particular cross-section, so that the load factor λ_a for the occurrence of alternating plasticity at this section is given by

$$\lambda_a(\mathscr{M}^{\max}-\mathscr{M}^{\min}) = 2M_p \qquad (8.23)$$

Static Theorem of Incremental Collapse

The Static Theorem of incremental collapse is a statement of the Shake-down Theorem, excluding the conditions (8.23) which refer to alternating plasticity, and is as follows:

"If there exists any distribution of residual bending moments throughout a frame to which the maximum and minimum elastic moments corresponding to a set of variable repeated load limits λW can be added without exceeding the fully plastic moment at any section, the load factor λ cannot be greater than the incremental collapse load factor λ_s."

Kinematic Theorem of Incremental Collapse

It was shown in Section 8.2 that a value of λ can be found to correspond to any assumed mechanism of incremental collapse. The Kinematic Theorem states that

"The load factor λ which is found to correspond to any assumed incremental collapse mechanism cannot be less than the incremental collapse load factor λ_s."

This theorem, which is the precise counterpart of the Kinematic Theorem of plastic collapse, can be proved directly (Neal, 1963), but like the plastic collapse theorem it can also be proved by a simple physical argument based on the Static Theorem. Imagine that the frame is strengthened by increasing the fully plastic moment indefinitely at every cross-section except where the hinges occur in the assumed mechanism. This strengthening is achieved by increasing the fully plastic moment indefinitely, while leaving the elastic properties of the members unaltered. By the Static Theorem, the incremental collapse load factor of the frame could not thereby be reduced, since the inequalities (8.15) would continue to be satisfied. The load factor λ, computed from the assumed incremental collapse mechanism, would of necessity be the actual incremental collapse load factor for this frame, and could not therefore be less than λ_s.

Uniqueness theorem of Incremental Collapse

This theorem, like the corresponding theorem for plastic collapse, is merely a combination of the Static and Kinematic Theorems, and is as follows:

"If there exists at least one distribution of residual bending moments throughout a frame, such that when the maximum and minimum elastic bending moments corresponding to a set of variable repeated load limits λW are added the fully plastic moment is never exceeded, but is attained at enough cross-sections to cause failure of the frame by incremental collapse, then the load factor λ is equal to the incremental collapse load factor λ_s."

8.6. TRIAL AND ERROR METHOD FOR DETERMINING λ_s

The use of the incremental collapse theorems can be illustrated by considering a simple example of the trial and error method for determining λ_s. In this method, a mechanism of incremental collapse is assumed, and the corresponding values of the load factor and of the residual bending moments are determined. The maximum and minimum elastic bending moments for the value λ' of the load factor thus obtained are then added to the residual bending moments to give the resulting maximum and minimum values of bending moment corresponding to the assumed mechanism.

If none of these peak bending moments exceeds the fully plastic moment, all the requirements of the Uniqueness Theorem have been met, so that λ' must be equal to the incremental collapse load factor λ_s. If, however, one or more of the peak values of the bending moment exceeds the fully plastic moment in magnitude, the assumed mechanism cannot be the actual incremental collapse mechanism. In this case the process must then be repeated for another assumed mechanism, the choice of this mechanism being guided by the results of the first trial.

The example which will be considered is the frame of Fig. 8.1(a). It will now be supposed that there is a vertical load V at C which can vary between the limits $(\lambda W, -\lambda W)$, and that there is also a horizontal load H at D which can vary between the limits $(2\lambda W, 0)$.

The first step in the analysis is to determine the elastic bending moments in the frame due to the loads $V = W$ and $H = 2W$ acting independently. These are given in Table 8.1, which also shows the values of $\mathscr{M}^{\max}$ and $\mathscr{M}^{\min}$ corresponding to the load limits $(W, -W)$ for V and $(2W, 0)$ for H.

TABLE 8.1

Cross-section	A	B	C	D	E
$V = W$	$0{\cdot}100WL$	$-0{\cdot}200WL$	$0{\cdot}300WL$	$-0{\cdot}200WL$	$0{\cdot}100WL$
$H = 2W$	$-0{\cdot}625WL$	$0{\cdot}375WL$	0	$-0{\cdot}375WL$	$0{\cdot}625WL$
$\mathscr{M}^{\max}$	$0{\cdot}100WL$	$0{\cdot}575WL$	$0{\cdot}300WL$	$0{\cdot}200WL$	$0{\cdot}725WL$
$\mathscr{M}^{\min}$	$-0{\cdot}725WL$	$-0{\cdot}200WL$	$-0{\cdot}300WL$	$-0{\cdot}575WL$	$-0{\cdot}100WL$

A further preliminary is to establish the equations of equilibrium which must be obeyed by the residual bending moments. These may be obtained by the virtual work method which was explained in Section 3.3 of Chapter 3. This involves using the residual bending moments m in the virtual work equation, in conjunction with the deformations of any suitable mechanism. For this purpose the three mechanisms of Figs. 8.1(b), (c) and (d) may be used in turn.

The equation of equilibrium corresponding to the mechanism of Fig. 8.1(b) was in fact derived in Section 8.2, equation (8.5), by this method, and is

$$-m_A + 2m_C - 2m_D + m_E = 0 \tag{8.5}$$

The equations corresponding to the mechanisms of Figs. 8.1(c) and (d) are similarly found to be

$$-m_B+2m_C-m_D = 0 \tag{8.24}$$

$$-m_A+m_B-m_D+m_E = 0 \tag{8.25}$$

It should be noted in passing that these three equations are not independent; equation (8.5) is obtained if the other two equations are added together.

First trial

Suppose now that it is first assumed that the incremental collapse mechanism is the mechanism of Fig. 8.1(b), with hinges at A, C, D and E. The corresponding residual bending moments at these four cross-sections can at once be deduced. For instance, the bending moment corresponding to the plastic hinge at A is $-M_p$, and this value must just be attained when the minimum elastic bending moment at A is added to the residual bending moment. With a load factor λ', it follows that

$$m_A+\lambda'\mathscr{M}^{\min} = -M_p,$$

or

$$m_A-0{\cdot}725\lambda' WL = -M_p, \tag{8.26}$$

using the value of $\mathscr{M}^{\min}$ given in Table 8.1. At the other hinge positions, it is found that

$$m_C+0{\cdot}300\lambda' WL = M_p \tag{8.27}$$

$$m_D-0{\cdot}575\lambda' WL = -M_p \tag{8.28}$$

$$m_E+0{\cdot}725\lambda' WL = M_p \tag{8.29}$$

These four residual bending moments are now substituted in equation (8.5), the equilibrium equation derived from the mechanism of Fig. 8.1(b). This gives

$$-(0{\cdot}725\lambda' WL-M_p)+2(M_p-0{\cdot}300\lambda' WL)$$
$$-2(0{\cdot}575\lambda' WL-M_p)+(M_p-0{\cdot}725\lambda' WL) = 0,$$

or

$$3{\cdot}2\lambda' WL = 6M_p,$$

so that

$$\lambda' = 1.875M_p/WL$$

With this value of λ', the residual bending moments m_A, m_C, m_D and m_E are at once found from equations (8.26)–(8.29). The value of m_B is then determined from either equation (8.24) or (8.25). These residual bending moments are given in Table 8.2, together with the values of the maximum and minimum elastic bending moments $\lambda'\mathscr{M}^{max}$ and $\lambda'\mathscr{M}^{min}$. The corresponding values of the maximum and minimum bending moments, $M^{max} = m+\lambda'\mathscr{M}^{max}$ and $M^{min} = m+\lambda'\mathscr{M}^{min}$, are also tabulated.

TABLE 8.2

Mechanism of Fig. 8.1(b). $\lambda' = 1{\cdot}875M_p/WL$

Cross-section	A	B	C	D	E
m	$0{\cdot}359M_p$	$0{\cdot}797M_p$	$0{\cdot}438M_p$	$0{\cdot}078M_p$	$-0{\cdot}359M_p$
$\lambda'\mathscr{M}^{max}$	$0{\cdot}188M_p$	$1{\cdot}078M_p$	$0{\cdot}562M_p$	$0{\cdot}375M_p$	$1{\cdot}359M_p$
$\lambda'\mathscr{M}^{min}$	$-1{\cdot}359M_p$	$-0{\cdot}375M_p$	$-0{\cdot}562M_p$	$-1{\cdot}078M_p$	$-0{\cdot}188M_p$
M^{max}	$0{\cdot}547M_p$	$1{\cdot}875M_p$	M_p	$0{\cdot}453M_p$	M_p
M^{min}	$-M_p$	$0{\cdot}422M_p$	$-0{\cdot}124M_p$	$-M_p$	$-0{\cdot}547M_p$

It will be seen that the value of M^{max} at B is $1{\cdot}875M_p$, and so the assumed mechanism cannot be the incremental collapse mechanism, since one of the requirements of the Uniqueness Theorem is violated. From the Kinematic Theorem, it follows that λ' must be an upper bound on the value of λ_s, so that

$$\lambda_s < 1{\cdot}875M_p/WL$$

A lower bound on the value of λ_s can also be derived from the results of this analysis, the process being similar to that employed in Section 8.4 to obtain a lower bound on λ_c. If all the entries in Table 8.2, together with the value of λ', were reduced in the ratio 1 : 1·875, a distribution of residual bending moments and a set of maximum and minimum elastic bending moments would be obtained such that none of the values of M^{max} and M^{min} exceeded M_p in magnitude. The requirements of the Static Theorem would therefore be met, and the corresponding value of λ, namely M_p/WL, would therefore be a lower bound on λ_s. Combining this with the upper bound just obtained,

$$M_p/WL < \lambda_s < 1{\cdot}875 M_p/WL$$

Second trial

These bounds are too far apart to be of use, and so a second trial must be made. The previous calculation gave a value of M^{max} of $1{\cdot}875 M_p$ at B, and so the mechanism to be assumed should involve a plastic hinge of positive sign at this cross-section. Thus the sidesway mechanism of Fig. 8.1(d) is chosen rather than the beam mechanism of Fig. 8.1(c), for this latter mechanism has a negative hinge rotation at B.

The calculation follows the same pattern as before, and details need not be given. The results are summarised in Table 8.3, the corresponding load factor λ'' being $1{\cdot}538 M_p/WL$.

TABLE 8.3

Mechanism of Fig. 8.1(d). $\lambda'' = 1{\cdot}538 M_p/WL$

Cross-section	A	B	C	D	E
m	$0{\cdot}115M_p$	$0{\cdot}115M_p$	0	$-0{\cdot}115M_p$	$-0{\cdot}115M_p$
$\lambda''\mathscr{M}^{max}$	$0{\cdot}154M_p$	$0{\cdot}885M_p$	$0{\cdot}462M_p$	$0{\cdot}308M_p$	$1{\cdot}115M_p$
$\lambda''\mathscr{M}^{min}$	$-1{\cdot}115M_p$	$-0{\cdot}308M_p$	$-0{\cdot}462M_p$	$-0{\cdot}885M_p$	$-0{\cdot}154M_p$
M^{max}	$0{\cdot}269M_p$	M_p	$0{\cdot}462M_p$	$0{\cdot}193M_p$	M_p
M^{min}	$-M_p$	$-0{\cdot}193M_p$	$-0{\cdot}462M_p$	$-M_p$	$-0{\cdot}269M_p$

It will be seen that none of the values of M^{max} and M^{min} exceeds M_p in magnitude. The requirements of the Uniqueness Theorem are therefore met, and so the value of λ_s is $1{\cdot}538 M_p/WL$.

Appendix A
Proof of principle of virtual work for plane frames

Consider first a typical member ij of the frame. The member is supposed to be initially straight and lying in a direction OT, which makes an angle α_{ij} with the horizontal. Figure A.1(a) shows the distorted member, with an origin O taken at the datum position of the end i, an axis OT coinciding with the datum direction of the undistorted member, and an axis ON perpendicular to OT in the plane of the frame.

The curvature κ of the member is taken as positive when in the sense shown in Fig. A.1(b), and the change of slope ψ as positive when in the sense of either of the clockwise end rotations which are shown as ψ_{ij} and ψ_{ji}. The geometrical relations between κ, ψ and the transverse deflection y_n in the direction ON are then

$$\psi = \frac{\mathrm{d}y_n}{\mathrm{d}s} \tag{A.1}$$

$$\kappa = \frac{\mathrm{d}\psi}{\mathrm{d}s}, \tag{A.2}$$

it being assumed that the deflections are small enough for these relations to hold true. δs represents an element of length of the member in the direction OT.

In addition, the axial tensile strain ε in the member is related to the axial displacement y_t in the direction OT by the equation

$$\varepsilon = \frac{\mathrm{d}y_t}{\mathrm{d}s} \tag{A.3}$$

The sign conventions for shear force F, bending moment M, axial tension P, and the normal and tangential intensities of loading w_n and w_t at any section of the member, are defined

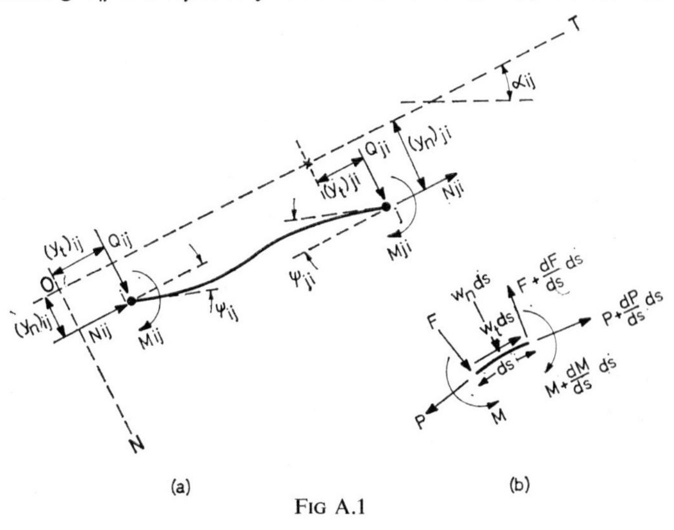

FIG A.1

in Fig. A.1(b). It will, however, be seen from Fig. A.1(a) that the end moments M_{ij} and M_{ji} are both defined as positive when acting clockwise on the member. It follows that

$$\left.\begin{aligned} \text{at } i, \quad & M = -M_{ij} \\ \text{at } j, \quad & M = M_{ji} \end{aligned}\right\} \tag{A.4}$$

Further, Fig. A.1(a) shows that the transverse forces Q_{ij} and Q_{ji} at the ends of the member are both defined as positive when acting in the direction ON, so that

$$\left.\begin{aligned} \text{at } i, \quad & F = Q_{ij} \\ \text{at } j, \quad & F = -Q_{ji} \end{aligned}\right\} \tag{A.5}$$

The normal forces N_{ij} and N_{ji} at the ends of the member are defined as positive when acting in the direction OT, so that

$$\left.\begin{aligned} \text{at } i, \quad P &= -N_{ij} \\ \text{at } j, \quad P &= N_{ji} \end{aligned}\right\} \tag{A.6}$$

The equilibrium relations for the element of the member depicted in Fig. A.1(b) are

$$F = \frac{\mathrm{d}M}{\mathrm{d}s} \tag{A.7}$$

$$w_n = \frac{\mathrm{d}F}{\mathrm{d}s} \tag{A.8}$$

$$w_t = -\frac{\mathrm{d}P}{\mathrm{d}s} \tag{A.9}$$

The proof of the principle is based upon the geometrical relations given in equations (A.1), (A.2) and (A.3), together with the equilibrium relations of equations (A.7), (A.8) and (A.9). Further geometrical and equilibrium relations are also required and will be given later.

The first step is to consider the value of

$$\int_i^j (M\kappa + P\varepsilon)\,\mathrm{d}s,$$

where M and P are distributions of bending moment and axial force in the member which satisfy the requirements of equilibrium, and κ and ε are distributions of curvature and axial strain satisfying the requirements of compatibility.

Using equation (A.2), it follows by integrating by parts that

$$\begin{aligned} \int_i^j M\kappa\,\mathrm{d}s &= \int_i^j M\frac{\mathrm{d}\psi}{\mathrm{d}s}\,\mathrm{d}s \\ &= [M\psi]_i^j - \int_i^j \frac{\mathrm{d}M}{\mathrm{d}s}\psi\,\mathrm{d}s \\ &= [M_{ji}\psi_{ji} + M_{ij}\psi_{ij}] - \int_i^j \frac{\mathrm{d}M}{\mathrm{d}s}\psi\,\mathrm{d}s, \end{aligned} \tag{A.10}$$

making use of equations (A.4).

From equations (A.1) and (A.7), and integrating by parts again, it is found that

$$\int_i^j \frac{dM}{ds}\,\psi\,ds = \int_i^j F\,\frac{dy_n}{ds}\,ds$$

$$= [Fy_n]_i^j - \int_i^j \frac{dF}{ds}\,y_n\,ds$$

$$= -[Q_{ji}(y_n)_{ji} + Q_{ij}(y_n)_{ij}] - \int_i^j w_n y_n\,ds \qquad \text{(A.11)}$$

using equations (A.5) and also the equilibrium relation equation (A.8).

Combining equations (A.10) and (A.11), it follows that

$$\int_i^j M\kappa\,ds = [M_{ji}\psi_{ji} + M_{ij}\psi_{ij}]$$

$$+[Q_{ji}(y_n)_{ji} + Q_{ij}(y_n)_{ij}] + \int_i^j w_n y_n\,ds \qquad \text{(A.12)}$$

Similarly, using equations (A.3) and (A.9), and integrating by parts, it follows that

$$\int_i^j P\varepsilon\,ds = \int_i^j P\,\frac{dy_t}{ds}\,ds$$

$$= [Py_t]_i^j - \int_i^j \frac{dP}{ds}\,y_t\,ds$$

$$= [N_{ji}(y_t)_{ji} + N_{ij}(y_t)_{ij}] + \int_i^j w_t y_t\,ds, \qquad \text{(A.13)}$$

making use of equations (A.6) and also the equilibrium equation (A.9).

Equations (A.12) and (A.13) are now summed to cover all the members of the frame. The terms

$$\int_i^j M\kappa\,ds, \qquad \int_i^j P\varepsilon\,ds, \qquad \int_i^j w_n y_n\,ds \quad \text{and} \quad \int_i^j w_t y_t\,ds,$$

which are integrations over the length of the member *ij*, present no difficulty; the integrations merely have to extend over all the members of the frame, and are then denoted by $\oint$. The remaining terms, such as $[M_{ji}\psi_{ji}+M_{ij}\psi_{ij}]$, refer to both ends of the member *ij*; a summation over all the members therefore covers each end of each member of the frame. This can also be achieved by summing first over all the members radiating from a particular joint, and then over all the joints, as follows:

$$\sum_{\text{bars}}[M_{ji}\psi_{ji}+M_{ij}\psi_{ij}] = \sum_{\text{joints}}\sum_{j} M_{ij}\psi_{ij}, \tag{A.14}$$

where $\sum_{j}$ denotes a summation over all the bars radiating from the joint *i*. Using equation (A.14), together with equivalent results, equations (A.12) and (A.13) are now added together and summed over all the bars of the frame to give

$$\oint (M\kappa+P\varepsilon)\,\mathrm{ds} = \oint (w_n y_n+w_t y_t)\,\mathrm{ds} + \sum_{\text{joints}}\sum_{j}[M_{ij}\psi_{ij}+Q_{ij}(y_n)_{ij}+N_{ij}(y_t)_{ij}] \tag{A.15}$$

Further consideration is now given to the terms $Q_{ij}(y_n)_{ij}$ and $N_{ij}(y_t)_{ij}$ in equation (A.15) by taking into account the

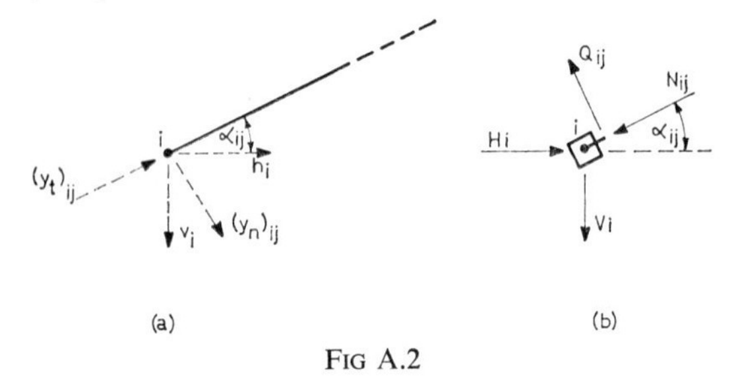

FIG A.2

conditions of joint equilibrium and compatibility of joint displacements. The end displacements $(y_n)_{ij}$ and $(y_t)_{ij}$ are related to the horizontal and vertical components of deflection

h_i and v_i of the joint i, which are defined in Fig. A.2(a), as follows:

$$(y_n)_{ij} = h_i \sin \alpha_{ij} + v_i \cos \alpha_{ij} \tag{A.16}$$

$$(y_t)_{ij} = h_i \cos \alpha_{ij} - v_i \sin \alpha_{ij} \tag{A.17}$$

Further, if there is an external load applied to joint i, with horizontal and vertical components H_i and V_i, as defined in Fig. A.2(b), the equations of equilibrium for this joint are

$$H_i = \sum_j (N_{ij} \cos \alpha_{ij} + Q_{ij} \sin \alpha_{ij}) \tag{A.18}$$

$$V_i = \sum_j (-N_{ij} \sin \alpha_{ij} + Q_{ij} \cos \alpha_{ij}) \tag{A.19}$$

Multiplying equation (A.18) by h_i and equation (A.19) by v_i, and adding, it is found that

$$H_i h_i + V_i v_i = \sum_j [N_{ij}(h_i \cos \alpha_{ij} - v_i \sin \alpha_{ij}) + Q_{ij}(h_i \sin \alpha_{ij} + v_i \cos \alpha_{ij})]$$

Using equations (A.16) and (A.17), this becomes

$$H_i h_i + V_i v_i = \sum_j [N_{ij}(y_t)_{ij} + Q_{ij}(y_n)_{ij}]$$

Summing this equation over all the joints, it follows that

$$\sum_{\text{joints}} (H_i h_i + V_i v_i) = \sum_{\text{joints}} \sum_j [N_{ij}(y_t)_{ij} + Q_{ij}(y_n)_{ij}] \tag{A.20}$$

Combining this result with equation (A.15), it follows that

$$\oint (M\kappa + P\varepsilon)\,ds = \oint (w_n y_n + w_t y_t)\,ds + \sum_{\text{joints}} (H_i h_i + V_i v_i) + \sum_{\text{joints}} \sum_j M_{ij}\psi_{ij} \tag{A.21}$$

Finally, it is necessary to interpret the term involving $M_{ij}\psi_{ij}$ in equation (A.21). Figure A.3 shows conditions at a typical joint i. The rotation of the member ij is ψ_{ij} clockwise, whereas the joint itself is supposed to have rotated clockwise by ϕ_i.

There is thus a hinge rotation θ_{ij} between the member and the joint, which is given by

$$\theta_{ij} = \phi_i - \psi_{ij} \tag{A.22}$$

The member ij exerts a counter-clockwise moment M_{ij} on the joint, and if there is a clockwise external couple C_i applied to the joint the condition for rotational equilibrium is

$$C_i = \sum_j M_{ij}$$

It follows at once that

$$C_i\phi_i = \sum_j M_{ij}\phi_i,$$

and from equation (A.22),

$$C_i\phi_i = \sum_j M_{ij}(\theta_{ij} + \psi_{ij}) \tag{A.23}$$

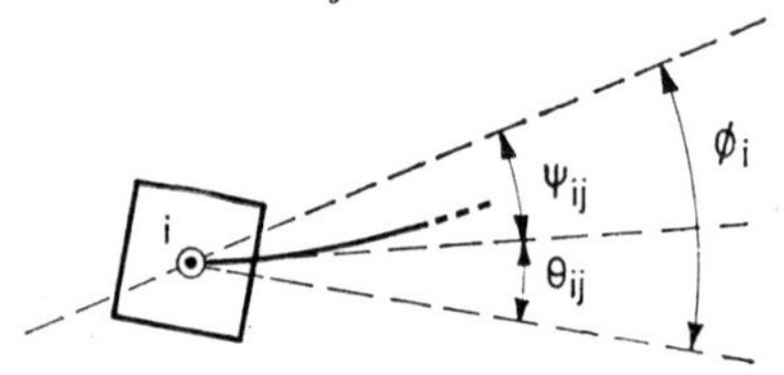

FIG A.3

Summing equation (A.23) over all the joints, and rearranging,

$$\sum_{\text{joints}} \sum_j M_{ij}\psi_{ij} = \sum_{\text{joints}} C_i\phi_i - \sum_{\text{joints}} \sum_j M_{ij}\theta_{ij} \tag{A.24}$$

The last term is summed over each end of all the members, and therefore covers all possible hinge positions, so that equation (A.24) may be written as

$$\sum_{\text{joints}} \sum_j M_{ij}\psi_{ij} = \sum_{\text{joints}} C_i\phi_i - \sum_{\text{hinges}} M_{ij}\theta_{ij} \tag{A.25}$$

Combining equation (A.25) with equation (A.21), and rearranging, the virtual work equation is obtained in the form

$$\sum_{\text{joints}} (H_i h_i + V_i v_i) + \oint (w_n y_n + w_t y_t)\, ds + \sum_{\text{joints}} C_i\phi_i$$
$$= \oint (M\kappa + P\varepsilon)\, ds + \sum_{\text{hinges}} M_{ij}\theta_{ij} \tag{A.26}$$

A study of this equation reveals that all the terms on the left-hand side represent virtual work done by external loads and couples, whereas all the terms on the right-hand side represent virtual work absorbed in the members and hinges. Thus the sign conventions are no longer of importance, so long as there is consistency; for instance, if sagging bending moments are regarded as positive then sagging curvatures must also be taken as positive.

For convenience of reference the suffices may now be discarded, and the final form of the equation is

$$\sum_{\text{joints}} (H^*h^{**}+V^*v^{**}+C^*\phi^{**})+\oint (w_n^*\, y_n^{**}+w_t^*\, y_t^{**})\,\mathrm{d}s$$

$$= \sum_{\text{hinges}} M^*\theta^{**}+\oint (M^*\kappa^{**}+P^*\varepsilon^{**})\,\mathrm{d}s \quad \text{(A.27)}$$

In this equation a single asterisk is used to denote each component of the force system, and a double asterisk is used to denote each component of the deformation system. This serves as a reminder that equation (A.27) is true provided only that the force system satisfies all the requirements of equilibrium, and the deformation system satisfies all the requirements of compatibility. The force and deformation systems need not be related as cause and effect, and indeed either or both of the systems can be hypothetical, rather than actual systems occurring in any particular problem.

There is little difficulty in extending the proof to cover the case of initially curved members, but details will not be given here; the only difference is that κ becomes the change of curvature.

Bibliography

ARGYRIS, J. H. and KELSEY, S. *Energy Theorems and Structural Analysis*. Butterworths (1960).

BETTI, *Nuovo Cimento* (2), Nos. 7, 8 (1872).

BLEICH, H. Über die Bemessung statisch unbestimmter Stahltragwerke unter Berücksichtigung des elastischplastischen Verhaltens des Baustoffes. *Bauingenieur*, **13**, 261 (1932).

CASTIGLIANO, A. *Théorie de l'Equilibre des Systèmes Élastiques et ses Applications*. Turin (1879). Translation by E. S. ANDREWS, *Elastic Stresses in Structures*. Scott, Greenwood; London (1919).

CHARLTON, T. M. The concepts of real and virtual work. *Engineering*, **180**, 139 (1955).

—— Statically indeterminate frames: the two basic approaches to analysis. *Engineering*, **182**, 822 (1956).

CROSS, HARDY. Analysis of continuous frames by distributing fixed end moments. *Proc. Amer. Soc. Civ. Engrs.* (May 1930) and *Trans. Amer. Soc. Civ. Engrs.*, **96**, 1 (1932).

ENGESSER, F. Ueber statisch unbestimmte Träger bei beliebigem Formänderungs—Gesetze und über den Satz von der kleinsten Ergänzungsarbeit. *Z. Arch. u. Ing. Verein Hannover*, **35** (1889).

FEINBERG, S. M. The principle of limiting stress (Russian) *Prikl. Mat. i Mekh.*, **12**, 63 (1948).

GREENBERG, H. J. The principle of limiting stress for structures. 2nd Symposium on Plasticity, Brown Univ., April 1949.

—— and PRAGER, W. On limit design of beams and frames. *Trans. Amer. Soc. Civ. Engrs.*, **117**, 447 (1952). (First published as Tech. Rep. Al8–1, Brown Univ. (1949)).

GVOZDEV, A. A. The determination of the value of the collapse load for statically indeterminate systems undergoing plastic deformation. *Proceedings of the Conference on Plastic Deformations*. December 1936, p. 19. Akademiia Nauk SSSR, Moscow–Leningrad (1938). Translated by R. M. HAYTHORNTHWAITE, *Int. J. Mech. Sci.*, **1**, 322 (1960).

HEYMAN, J. On the estimation of deflexions in elastic-plastic framed structures. *Proc. Instn. Civ. Engrs.*, **19**, 39 (1961).

—— *Beams and Framed Structures*. Pergamon (1964).

HOFF, N. J. *The Analysis of Structures*. Wiley, N.Y. (1956).

HORNE, M. R. Fundamental propositions in the plastic theory of structures. *J. Instn. Civ. Engrs.*, **34,** 174 (1950).

KIST, N. Leidt een Sterkteberekening, die Uitgaat van de Evenredigheid van Kracht en Vormverandering, tot een goede Constructie van Ijzeren Bruggen en gebouwen? Inaugural Dissertation, Polytechnic Institute, Delft (1917).

LAMB, E. H. The principle of virtual velocities and its application to the theory of elastic structures. Instn. Civ. Engrs., *Selected Engineering Papers*, No. 10 (1923).

LIVESLEY, R. K. *Matrix Methods of Structural Analysis*. Pergamon (1964).

MAIER-LEIBNITZ, H. Versuche mit eingespannten und einfachen Balken von I-form aus St 37. *Bautechnik*, **7**, 313 (1929).

MATHESON, J. A. L. *Hyperstatic Structures*, vol. I. Butterworths (1959).

MAXWELL, J. CLERK. On the calculation of the equilibrium and stiffness of frames. *Phil. Mag.* Series 4, **27**, 294 (1864). See also a commentary on this paper by Niles (1950).

MELAN, E. Theorie statisch unbestimmter Systeme. Prelim. Pubn. 2nd Congr. Intern. Assn. Bridge and Struct. Engng., 43, Berlin (1936).

—— Die Theorie statisch unbestimmter Systeme aus ideal plastischen Baustoff. *S. B. Akad. Wiss. Wien.* (Abt. IIa), **145,** 195 (1936).

MOHR, O. Beitrag zur Theorie der Bogenfachwerksträger. *Z. Arch. u. Ing. Verein Hannover*, p. 223 (1874).

—— Beitrage zur Theorie des Fachwerks. *Z. Arch. u. Ing. Verein Hannover*, p. 509 (1874), p. 17 (1875).

—— *Technische Mechanik*. Ingenieur-Verein am Polytechnicum zu Stuttgart (1877).

—— Beitrag zur Theorie des Bogenfachwerks. *Z. Arch. u. Ing. Verein Hannover*, p. 243 (1881).

—— Ueber das sogenannte Prinzip der kleinste Deformationsarbeit. Wochenblatt für Architekten und Ingenieure, p. 171 (1883).

—— Beitrag zur Theorie des Fachwerks. *Civilingenieur*, p. 289 (1885).

—— Ueber Geschwindigkeitsplane und Beschleunigungsplane. *Civilingenieur*, p. 631 (1887).

—— Die Berechnung der Fachwerke mit starren Knotenverbindungen. *Civilingenieur*, p. 577 (1892), p. 67 (1893).

—— *Abhandlung aus dem Gebiete der Technischen Mechanik* (1906).

MÜLLER-BRESLAU, H. Der Satz von der Abgeleiteten der ideellen Formänderungs-Arbeit. *Z. Arch. u. Ing. Verein Hannover*, **30** (1884).

—— *Die neueren Methoden der Festigkeitslehre und der Statik der Baukonstruktionen*. Körner, Leipzig (1886).

NEAL, B. G. The behaviour of framed structures under repeated loading. *Quart. J. Mech. Appl. Math.*, **4,** 78 (1951).

—— Virtual work and the moment distribution method. *Engineering*, **183,** 47 (1957).

NEAL, B. G. *The Plastic Methods of Structural Analysis*. Chapman & Hall (London), Wiley (N.Y.) (1963).

NILES, A. S. The "Energy Method", which one? *J. Eng. Education*, **33**, 698 (1943).

—— Clerk Maxwell and the theory of indeterminate structures. *Engineering*, **170**, 194 (1950).

OSTENFELD, A. *Die Deformationsmethode*. Springer, Berlin (1926).

PARKES, E. W. *Braced Frameworks*. Pergamon (1964).

PRAGER, W. *Introduction to Mechanics of Continua*. Ginn (1961) Ch. VIII.

LORD RAYLEIGH. A statical theorem. *Phil. Mag.*, **48**, 452 (1874); **49**, 183 (1875).

SYMONDS, P. S. and PRAGER, W. Elastic-plastic analysis of structures subjected to loads varying arbitrarily between prescribed limits. *J. Appl. Mech.*, **17**, 315 (1950).

Recommended further reading

ARGYRIS, J. H. and KELSEY, S. *Energy Theorems and Structural Analysis.* Butterworths (1960).

BROWN, E. H. The energy theorems of structural analysis. *Engineering*, **179**, 305 (1955).

CHARLTON, T. M. Some notes on the analysis of redundant systems by means of the concept of conservation of energy. *J. Franklin Inst.*, **250**, 543 (1950).

—— Analysis of statically indeterminate structures by the complementary energy method. *Engineering*, **174**, 389 (1952).

—— Strain compatibility conditions of grossly distorted structures by virtual work. *Civil Engng.*, **58**, 325 (1963).

HOFF, N. J. *The Analysis of Structures.* Wiley, N.Y. (1956).

KING, J. W. H. Some notes on plane frames not obeying Hooke's Law. *Engineer*, **196**, 4 (1953).

LAMB, E. H. The principle of virtual velocities and its application to the theory of elastic structures. Instn. Civ. Engrs., *Selected Engineering Papers*, No. 10 (1923).

MATHESON, J. A. L. Castigliano's "Theorem of Compatibility". *Engineering*, **180**, 828 (1955).

MATHESON, J. A. L. *Hyperstatic Structures*, vol. I. Butterworths (1959).

PIPPARD, A. J. S. *Strain Energy Methods of Stress Analysis.* Longmans, Green (1928).

SOUTHWELL, R. V. *An Introduction to the Theory of Elasticity.* Oxford (1936). (This book contains interesting derivations of Castigliano's theorems.)

TIMOSHENKO, S. *History of Strength of Materials.* McGraw-Hill (1953).

WESTERGAARD, H. On the method of complementary energy. *Trans. Amer. Soc. Civ. Engrs.*, **107**, 765 (1942).

WILLIAMS, D. The relations between the energy theorems applicable in structural theory. *Phil. Mag.*, **26**, 617 (1938).

—— The use of the principle of minimum potential energy in problems of static equilibrium. Aer. Res. Cttee. R. and M. No. 1827 (1938).

Index

Alternating plasticity 176
Arch, influence lines for, *see* Influence lines
  redundancies 32–35
  superposition of forces in 22
  temperature stresses in 84–88

Basic truss, *see* Truss
Beam
  collapse load factor of 169–72
  deflections of, *see* Deflection calculations
  influence lines for, *see* Influence lines
  redundancies 8–10, 30–32
  superposition of forces in 21–22
Betti's Reciprocal Theorem 144–6
  application to model analysis 150–2

Cable carrying concentrated load 5–6
Castigliano's Theorem (Part I) 131–2, 140
Castigliano's Theorem (Part II) 93, 102–3
  determination of deflections by 106–7
Castigliano's Theorem of Compatibility 77–79
  derivation of compatibility equations by 83–84, 91–92
Compatibility, Castigliano's Theorem of, *see* Castigliano's Theorem of Compatibility
  Engesser's Theorem of, *see* Engesser's Theorem of Compatibility
Compatibility equations, derivation of, *see* Castigliano's Theorem of Compatibility; Engesser's Theorem of Compatibility, *and* Virtual Work
Compatibility method 10–15, 20, 58–92
Complementary Energy 54–56, 64, 71, 72, 87–88
  minimum of 73–74
  First Theorem of 93, 96–97, 101–2
Cyclic loading, *see* Loading

Datum geometry 28, 50, 54, 57, 66, 74, 84, 97, 103, 123, 126
Deflection calculations
  energy theorems for 101–3, 106–7
  for beam at plastic collapse 108–12
  for elastic beam 104–8, 149
  for statically determinate truss 43–46, 94–97

Deflection Calculations *cont.*
  for statically indeterminate truss 97–101
Dummy unit load 44, 46, 93, 95, 99, 149

Engesser's Theorem of Compatibility 71–73
  derivation of compatibility equations by 64–66, 70–71, 87–88
Equilibrium equations, derivation of, *see* Potential Energy *and* Virtual Work
Equilibrium method 10–15, 24, 113–40

Feinberg's axiom 168, 169
First Theorem of Complementary Energy, *see* Complementary Energy
First Theorem of Minimum Strain Energy, *see* Strain Energy
Flexibility matrix 69–70, 119
Flexible supports, *see* Supports
Free bending moment 21
Fully plastic moment 153–4

Gross deformations, exclusion of 4–7, 17, 22, 25, 28, 37, 41

Incremental collapse 159–62
  Kinematic Theorem of 164, 177
  mechanism of 161, 178–82
  Static Theorem of 177
  Uniqueness Theorem of 178
Incremental collapse load, calculation of 162–4
Incremental collapse load factor 172
  calculation of 178–82
  bounds on 181–2
Influence lines
  for arches 150–2
  for beams 49–50, 146–50

Kinematic Theorem
  of Incremental collapse, *see* Incremental collapse
  of Plastic collapse, *see* Plastic collapse

Lack of fit 74–77, 78, 123–5, 126
Linear elastic structures 25–29, 33–35, 78, 102–3, 106, 141, 144
Load factor
  of incremental collapse, *see* Incremental collapse
  of plastic collapse, *see* Plastic collapse
Loading
  cyclic 159
  proportional 155, 166, 169
  variable repeated 164, 172, 173, 177, 178

Maxwell's Reciprocal Theorem 142–4
  application to influence lines 146–50
Maxwell–Mohr equations 70
Maxwell–Mohr method for deflections 46, 96
Mechanism
  of incremental collapse, *see* Incremental collapse
  of plastic collapse, *see* Plastic collapse
Minimum
  of Complementary Energy, *see* Complementary Energy
  of Total Potential, *see* Total Potential
Moment distribution 49, 132, 140
Müller–Breslau's principle 147

Plastic collapse 155–7, 165–6
  Kinematic Theorem of 158, 168–9, 170–2
  mechanism of 157–8, 169–72

Static Theorem of 166–8, 169, 170, 171
Uniqueness Theorem of 169
Plastic collapse load, calculation of 158–9
Plastic collapse load factor 166–9
bounds on 169–72
Plastic hinge concept 153
Portal frame
analysis by compatibility method 79–84
analysis by equilibrium method 132–40
behaviour under cyclic loading 159–62
behaviour under proportional loading 155–7
equilibrium equations for 46–49
incremental collapse load of 162–4, 178–82
plastic collapse load of 158–9
redundancies 30–31
Potential Energy, Theorem of Minimum 125–9
derivation of equilibrium equations by 121–3, 140
Proportional loading, *see* Loading
Proving ring, redundancies 30–32

Reactant bending moment 21
Reciprocal Theorem, *see* Betti's Reciprocal Theorem *and* Maxwell's Reciprocal Theorem
Residual bending moment 161, 162–4, 173–5, 177, 178–82
Ring, loaded normal to plane 88–92

Safe bending moments 166
Shake-down 160–2
Theorem of 172–6
Sidesway 48, 182
Sinking of supports, *see* Supports
Skew-symmetrical loading 32–35
Slope-deflection equations 135

Static Theorem
of Incremental collapse, *see* Incremental collapse
of Plastic collapse, *see* Plastic collapse
Statically admissible bending moments 166
Stiffness, bounds on 130–1
Stiffness matrix 118–19
Strain Energy 50–54, 57, 78–79, 83–84, 91–92, 102–3, 106–7, 131–2
First Theorem of Minimum 132
Stress/strain relation, non-linear 51, 54, 55, 59, 73, 119, 129, 132
Structural analysis
basic conditions of 2–4
choice of variables for 10–14, 58, 66–67, 113, 114
direct and indirect approaches 14–15
of determinate and indeterminate structures 8–10
using models 150–2
Strut, eccentrically loaded 6–7
Superposition
for linear elastic structures 25–29, 33–35, 133
of displacements 22–25, 116, 126
of force systems 16–22, 61, 68, 75, 81, 101
Supports, flexible 52, 53, 56, 66, 72, 97, 127
sinking of 2, 3, 28, 32
Symmetry 30–35
Symmetry 30–35

Temperature strains 27, 50, 54, 66–71, 84–88, 94–101
Theorem of Minimum Potential Energy, *see* Potential Energy
Total Potential 57, 121–3, 127–9
Minimum of 129–30
Truss
analysis by compatibility method 11–12, 59–71, 74–77

Truss *cont.*
  analysis by equilibrium method 12–13, 114–25
  basic 20–21, 60, 62, 67, 78, 99
  determination of deflections 43–46, 94–101
  redundancies 30–34

Uniqueness Theorem
  of Incremental collapse, *see* Incremental collapse
  of Plastic collapse, *see* Plastic collapse

Variable repeated loading, *see* Loading

Virtual Work Principle 37–42, 183–90
  derivation of compatibility equations by 62–63, 68, 76–77, 81–82, 86, 90–91
  derivation of equilibrium equations by 46–49, 116–17, 124, 127, 134–9, 179
  determination of deflections by 43–46, 94–96, 97–101, 104–6, 107–12
  transformations 15, 36, 42–50, 59, 114

Yield load 156